Preface

This book forms part of the Macmillan Secondary Mathematics course, which has been written to correspond to the National Criteria for GCSE Mathematics examinations, and to the Mathematics National Curriculum assessment and testing framework. In particular the X-stream books are for the use of pupils who intend going on to the lower or intermediate levels of the GCSE Mathematics examination.

Book 3X contains some revision of topics covered in Books 1 and 2 of the series, which should help consolidate this work, and further extend the skills which have already been used. Each chapter contains worked examples, and sets of graded exercises. In addition there are many mathematics investigations which can be used to supplement skills learnt, and offer extensions to the exercises. The three revision exercises will also provide for additional practice if needed.

In terms of the National Curriculum, the work covered in Book 3X corresponds with Levels 3 – 6, and covers both the profile components, attributing weight corresponding to the weightings given in the attainment targets. Examples used are in the context of 'using and applying mathematics', and are in realistic contexts of everyday life at home and at work, within today's multicultural society.

1. Money (1)

Odds and evens

In Book 1 we were shown how numbers can be divided into those which are odd, and those which are even.

>   Odd numbers: 1, 3, 5, 7, 9 …
>   Even numbers: 2, 4, 6, 8, 10 …

Which of the coins we use are odd? Which are even?

In any monetary system the coins available can be used to make up any amount of money easily. The biggest amount of money we need, without using £ coins, is 99p, because anything over 99p makes up to £1 or more.

Investigation A

Find out how many different ways there are of making 99p using the coins in our monetary system.

Investigation B

Find out how many different ways there are of making 58p from the coins we use, but using no more than 15 coins in all! Try the same exercise for different amounts of money. Which amounts can be made up in the least number of ways?

Investigation C

In each case write down whether you get an odd or an even number.

(*a*) Pick an odd number, then add another odd number to it.

(*b*) Pick an odd number, then add an even number to it.

(*c*) Pick an even number, then add another even number to it.

Try some more examples until you build up a set of rules, which should show that even numbers are easy to make, but to make up odd numbers we first need both odd and even. Next try the same idea with three numbers; start by writing down all the possible combinations of odd or even numbers.

Investigation D

Write down the minimum number of coins you would need to represent each of the following amounts of money: 20p, 21p, 22p, 23p, 24p, 25p ... Are there any coins we could easily do without? Would you recommend the introduction of any new coins to reduce the amount of change we have to carry around?

Prime numbers

A **prime number** is a number which can be divided exactly only by itself and the number 1. For example, 5 is a prime since 5 and 1 are the only numbers which can be divided into it, whereas 6 is not prime because 2 and 3 can be divided exactly into it. The number 1 itself is not a prime number.

Investigation E

Which amounts of money between 2p and 99p represent prime numbers?

EXERCISE 1.1

1 Which of the numbers below are (*a*) prime, (*b*) odd, (*c*) even?
5, 7, 10, 13, 15, 22, 27, 29, 33, 39, 44, 47, 50.

2 Write down as many prime numbers as you can which are even.

3 Write down all the prime numbers between 30 and 40.

4 Which of these odd numbers are prime: 51, 55, 63, 67, 79, 81, 85?

5 How many prime numbers are there between 100 and 110?

6 Roads sometimes have odd-numbered houses on one side, and even-numbered houses on the other side.

(*a*) What is the number of the 15th house on the odd side?

(*b*) What is the number of the 17th house on the even side?

(*c*) What is the number of the 20th house on the odd side?

(*d*) What is the number of the 20th house on the even side?

(*e*) What are the numbers of the houses two doors away on either side of number 83?

Factors

A **factor** is a number which divides exactly into another number.

Example 1
Write down the factors of the numbers (*a*) 8 (*b*) 9.

(*a*) 8 has factors 1, 2, 4, and 8 since all these divide into 8 exactly.
(*b*) 9 has factors 1, 3, and 9.

EXERCISE 1.2

Find all the factors of the following numbers.

1 4   **2** 14   **3** 15   **4** 12   **5** 16   **6** 20   **7** 24

8 28   **9** 30   **10** 56   **11** 72   **12** 99

13 Write down the factors of each of the coins we use.

Common factors

A **common factor** is a factor of two or more numbers.

Example 2
Write down the common factors of 6 and 9.

6 has the factors 1, 2, 3, 6.
9 has the factors 1, 3, 9.
So 6 and 9 have common factors of 1 and 3.

Investigation F
Write down three prime numbers and find the factors of each. What do you notice? Do they share any common factors? Find out if there is anything you can say about factors of prime numbers.

EXERCISE 1.3

Find all the common factors of the numbers in each question.

1 2, 4   **2** 6, 8   **3** 8, 12   **4** 5, 15

5 12, 24   **6** 9, 21   **7** 15, 25   **8** 2, 8, 10

9 6, 18, 24   **10** 12, 16, 28

11 What are the common factors of all the coins we use?

12 Find three pairs of numbers which have five or more common factors.

Multiples

The multiples of 2 are 2, 4, 6, 8 ..., i.e. the 2 times table.
The multiples of 4 are 4, 8, 12, 16 ..., i.e. the 4 times table.

Example 3

Write down all the multiples of 2 from the numbers 3, 4, 5, 6, 7.

The multiples of 2 from the list are 4 and 6, since 2 is a factor of these numbers.

EXERCISE 1.4

1 Write out the first six multiples of the numbers:

(*a*) 5   (*b*) 8   (*c*) 10   (*d*) 13   (*e*) 15

2 From the list of numbers 2, 4, 5, 7, 8, 10, 12, 15, 16, 18, write out the numbers which are:

(*a*) multiples of 2, (*b*) multiples of 4, (*c*) multiples of 5,
(*d*) multiples of 6.

3 Find the multiples of 7 between 20 and 60.

4 Find the multiples of 12 between 100 and 170.

Common multiples

A **common multiple** is a multiple of two or more numbers.

Example 4

Find the common multiples of 2 and 4.

The multiples of 2 are 2, 4, 6, 8, 10 …
The multiples of 4 are 4, 8, 12, 16 …
So the numbers 4 and 8 are common multiples of 2 and 4. Try to find some more common multiples of 2 and 4. There are many more.

Investigation G

Which of our coins have a common multiple?

EXERCISE 1.5

Find the first three common multiples for each of the numbers used in Exercise 1.3, questions 1 to 8.

Decimals and place value

Numbers can be broken down into parts:

$£345$ is the same as $(3 \times £100) + (4 \times £10) + (5 \times £1)$

When we want to show an amount of money which includes pence we have to use the decimal dot: £345.76.
What does the 7 represent?

£100	£10	£1	10p	1p
3	4	5	7	6

The table shows that the 7 is the same as 7×10p, and 10p is exactly $£\frac{1}{10}$, so the 7 represents $£\frac{7}{10}$.
The table shows that the 6 is the same as 6×1p, and 1p is exactly $£\frac{1}{100}$, so the 6 represents $£\frac{6}{100}$.

$$£345.76 = £345 + \tfrac{7}{10} + \tfrac{6}{100}$$
$$£345.42 = £345 + \tfrac{4}{10} + \tfrac{2}{100}$$

Which is the larger? £345.76 since we have $£\frac{3}{10}$ and $£\frac{4}{100}$ more than the other amount of money.

EXERCISE 1.6

$£2.43 = £2 + £\frac{4}{10} + £\frac{3}{100}$

1 Write out the following amounts of money in the same way:

(a) £1.34   (b) £2.79   (c) £3.48   (d) £4.56   (e) £5.92

(f) £45.63   (g) £30.03   (h) £12.01   (i) £0.99   (j) £401.10

2 Place the following amounts of money in order of size, smallest first:

(a) £1.32, £1.35   (b) £3.90, £3.85   (c) £2.19, £2.20

(d) £5.10, £5.01   (e) £56.19, £56.91   (f) £20.50, £102.05

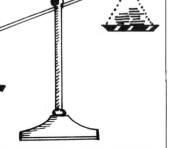

We have to remember that whatever amount of money we handle there should always be exactly two figures after the decimal point. Why? What is the maximum number of pence you can have which is less than £1? Add one penny on; what amount do you have now?

Try this sum on your calculator: £1.10 + £1.10 =
Copy the sum and write down what is displayed by the calculator. You should have 2.2.

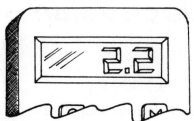

Describe this amount of money to your neighbour. How would you expect to see it written in a shop? How much would you pay for an article costing this much? Although the calculator has given us 2.2 as an answer, we should write it as £2.2<u>0</u> to give us two figures after the decimal dot. This amount is then £2 and 20 pence, quite clearly, and without confusion.

EXERCISE 1.7

Work out the following sums on your calculator, writing your answers clearly and correctly.

1 £3.25 + £1.65   **2** £2.37 + £2.73   **3** £5.82 + £6.68

4 £6.72 − £3.02   **5** £9.00 − £1.10   **6** £3.42 − £2.92

7 £2.65 + £8.14 + £6.01   **8** £5.58 + £9.21 + £5.11

9 £5.21 − £2.18 + £3.67   **10** £14.59 + £12.99 − £0.28

Write the following as a decimal amount of money, for example, $£\frac{3}{100} = £0.03$.

11 $£\frac{19}{100}$ **12** $£2\frac{7}{10}$ **13** $£1\frac{13}{100}$ **14** $£5\frac{1}{100}$

15 $£6\frac{8}{10}$ **16** $£7\frac{1}{10}$ **17** $£\frac{99}{100}$ **18** $£\frac{7}{100}$

19 $£8\frac{45}{100}$ **20** $£4\frac{4}{100}$

Investigation H

Using your calculator, find some amounts of money so that when we add them up we get an answer like £3.2, that is an answer with $£\frac{1}{10}$ and not $£\frac{1}{100}$. Add up the *last* digits of the amounts of money in each of your problems. What do you notice?

Decimals: addition and subtraction

Complete these sums on your calculator:

(*a*) £1+£3+£9   (*b*) 23p+42p+98p   (*c*) £2+28p+£4+32p

You should have answers of £13 and 163 pence for the first two, but if you have an answer of 66 for the last problem it shows you have misunderstood how to use your calculator. The last problem has a mixture of £ and pence: you must keep them separate on the calculator. We usually do this by writing everything in terms of £.

28p = £0.28   32p = £0.32

Now do (*c*) again: £2+£0.28+£4+£0.32 = £6.60
When setting this out as a sum on paper we should also make sure we
keep £ and pence separate

```
£  p
2.00   ← put in 00 for no pence
0.28
4.00   ← put in 00 for no pence
0.32
6.60
```

EXERCISE 1.8

Copy these six bills into your book. Set them out correctly as shown
above, and find the total amount.

1 £3.45+22p+78p+£5.50   **2** £45.12+£8.99+96p+£1.02

3 67p+78p+45p+£1.28   **4** £12.01+£6+6p+£3.45

5 £9.01+99p+£8+£1.10   **6** £7.80+85p+£2.90+£5+5p

Work out the cost of these bills:

7 Three compact discs.

8 Two pairs of roller skates.

9 Two comics and a box of chocolates.

10 One pair of shoes and a compact disc.

11 A pair of shoes and a pair of roller skates.

12 Four compact discs, two comics and a box of chocolates.

13 What would be my change from £20 if I bought a pair of roller
skates?

EXERCISE 1.9

Work out:

1 £32.96+£14.14+£6.02   **2** £34.18+£51.70+£95.12

3 £43.19+73p+£102.45   **4** 90p−37p

5 £13.93−£6.54   **6** £21−£4.60

7 £30.50−74p   **8** £100−£3.70

9 £2.50+£3.25−39p   **10** £7.87−£1.03+26p

11 £21+£15−15p

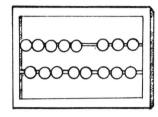

12 A pair of trousers were £17.90, and have been reduced by £1.80 in a sale. What is the sale price?

13 Susan bought four cards costing £1.05, 73p, 89p, £1.13. What change did she receive out of a £5 note?

14 Derek owes £200, and repays £40.10, £68.75 and £25.15. How much does he still owe?

15 The bill for two shirts was £9.60. If one shirt cost £4.70, what did the other shirt cost?

16 Omar bought three items costing £52.30, £46.81 and £120.56. What was the total cost?

17 Rajid buys a present costing £7.15 and pays for it with a £10 note. What change should he receive?

18 At a sale a bike costing £66.59 was sold for £55.99. How much was saved by buying the bike at the sale price?

EXERCISE 1.10

Find the total of each of these shopping bills.

1	2	3	4	5
0.32	0.44	0.13	0.05	0.13
0.12	0.11	0.14	0.14	0.13
0.12	0.06	0.14	0.08	0.19
0.12	0.19	0.14	0.12	0.15
0.12	0.54	0.45	0.12	0.23
0.34	0.45	0.23	0.16	0.16
1.45	0.14	0.45	0.65	0.34
1.32	0.35	0.65	2.57	1.76
0.18	0.34	0.19	0.28	0.29
0.56	0.67	0.76	0.82	0.54
0.56	0.73	0.43	0.33	0.54
0.56	0.45	0.43	0.38	0.54
0.56	0.20	0.30	2.23	0.22
5.70	0.20	0.36	0.20	0.46
4.99	0.20	0.45	0.32	2.85
0.34	0.78	1.87	2.63	1.12
0.04	0.37	0.74	1.74	0.69
8.54	0.73	0.73	0.38	0.85
0.56		0.99	0.54	0.76
0.54			1.05	0.22
			0.32	0.44
				0.56

Decimals: multiplication

Ten boxes of 'beef curry' costing £1.25 are to be bought. How much will they cost?

$$£1.25 \times 10 = £12.50$$

What has happened to the number?

Multiply these by ten: £1.03, £0.45, £33.33 and compare your answers. Can you see a quick way to multiply by ten without using a calculator?

When multiplying by ten all the digits move up one place, and we add on a nought to ensure we have two digits after the decimal point for the pence.

Suppose we wanted one hundred boxes costing £1.25

£1.25 × 100 = £125.00

What has happened to the number?

Multiply the other amounts of money by 100 and compare your answers again. All the digits have moved up two places. Perhaps you also noticed that all the answers end in '00', that is, they are all amounts in £ and no pence. Why do you think this is so?

Example 5

£3.05	£1.13
× 5	× 9
£15.25	£10.17

EXERCISE **1.11**

1 £3.25×10	**2** £1.01×10	**3** £7.90×10	**4** 15p×10
5 £3.25×100	**6** £1.01×100	**7** £7.90×100	**8** 15p×100
9 41p×5	**10** £4.30×8	**11** £6.81×7	**12** £2.66×5
13 £11.82×9	**14** £30.23×7	**15** £19.73×6	**16** £4.25×8
17 £5.09×12	**18** £32.13×13	**19** £27.50×48	**20** £39.99×65

EXERCISE **1.12**

Find the cost of:

Peas 45p/lb

Cauliflower 55p / lb

Carrots 28p/lb

Tomatoes 62p/ lb

Bananas 58p/ lb

Apples 32p/ lb

1 2 lb peas, 2 lb cauliflower and 1 lb carrots

2 2 lb bananas, 3 lb apples and 2 lb carrots

3 $1\frac{1}{2}$ lb carrots, 1 lb bananas and $\frac{1}{2}$ lb tomatoes

4 How much change from a £10 note would you want after buying 4 lb apples and $1\frac{1}{2}$ lb bananas?

5 Roger earns £738.53 salary in a month. How much should he earn in a year?

6 A firms pays part-time workers at £1.83 per hour. How much will they get for 11 hours' work?

7 A newspaper is issued on 294 days in a year. Find the total cost of buying the newspaper over the year when each copy is 35p.

8 LP records are reduced from £6.50 to £5.75. What saving will there be on buying six such records?

9 A series of text books is valued as follows: Bk1 £4.95, Bk2 £4.95, Bk3 £5.25, Bk4 £5.25, Bk5 £5.45. How much would it cost to buy 120 of each book in the series?

10 A firm pays its workers £2.45 per hour basic, plus £3.10 per hour overtime. What would be the total wage bill for a team of five staff who each worked a 35-hour basic week plus 5 hours' overtime?

Decimals: division

We buy a box of ten pens for £1.90. How much will each pen have cost?

$£1.90 \div 10 = £0.19$

What has happened to the number?
   We noticed when multiplying by ten that the digits moved to the left.
We can see that when we divide by ten the digits move to the right.
   Divide these by 10: £1.20, £45.60, £62.00

A similar process happens when we divide by 100.

$£19.00 \div 100 = £0.19$

What has happened to the numbers now?
   Try dividing these amounts by 100: £3.00, £12.00, £60.00
The digits move two places to the right. You might also notice that the amounts written in £ become amounts in pence. Why do you think this is?

Example 6

$$\begin{array}{r} £1.19 \\ 5 \overline{) £5.9^45} \end{array} \qquad \begin{array}{r} £0.97 \\ 12 \overline{) £11.6^84} \end{array}$$

EXERCISE 1.13

1 £51.00÷10	**2** £101.00÷10	**3** £20÷10	**4** £45÷10
5 £51.00÷100	**6** £101.00÷100	**7** £20÷100	**8** £45÷100
9 96p÷4	**10** £7.56÷6	**11** £77.36÷8	**12** £14.35÷7
13 £15.85÷5	**14** £10.08÷4	**15** £29.97÷9	**16** £6.88÷8
17 £19.08÷12	**18** £17.46÷18	**19** £245.70÷42	**20** £174÷58

EXERCISE 1.14

1 How many 5p coins can be exchanged for a £5 note?

2 Sheila is paid £82.25 for a 35-hour week. How much is this per hour?

3 A box of 36 bags of crisps costs £5.76. How much will each bag cost?

4 A group of twelve workers win £12 974.88 on the football pools. How much will each one receive?

5 A gas bill is £59.28 for one quarter (13 weeks). How much is this per week?

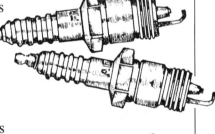

6 A crate of 1200 spark plugs is priced at £1176. A discount of £72 is given on the order. How much will it cost for *each* spark plug after deduction of the discount?

7 Amanda receives £1.40 pocket money a week. How much is this per day?

8 The 48 workers at a factory are given a bonus of £1352.16. How much will each receive?

9 Ball pens cost 48p each. How many could I buy for £5?

10 Vivek can buy four light bulbs at 63p each, or a pack of four for £2.44. How will he buy them, and what saving will he be making?

Directed numbers

Many people use banks and building societies for saving money, and some use them to borrow money. For example, if a bank account has £60 in it, and a cheque is written for £80 and cashed, then the account will be **overdrawn** by £20, that is, the account holder will owe the bank £20. How will the calculator show this? Work it out for yourself: $60-80 = ?$ Your calculator should show the answer as -20.
The minus sign tells you that the answer is not money in the bank account but money that is owed.

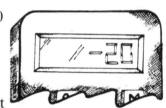

EXERCISE 1.15

For each problem you are told the amount of savings in the account, and the amount to be withdrawn. In each case find the amount owing.

1 Savings = £15   Withdrawal = £17

2 Savings = £23   Withdrawal = £26

3 Savings = £32   Withdrawal = £39

4 Savings = £9   Withdrawal = £11.50   **10** Savings = £10.01   Withdrawal = £11.10

5 Savings = £6   Withdrawal = £8.25   **11** Savings = £82.50   Withdrawal = £110.23

6 Savings = £5.50   Withdrawal = £7.25   **12** Savings = £15.37   Withdrawal = £55.27

7 Savings = £3.15   Withdrawal = £4.32   **13** Savings = £39.82   Withdrawal = £40.10

8 Savings = £12.71   Withdrawal = £15.97   **14** Savings = £29.30   Withdrawal = £35.27

9 Savings = £85.31   Withdrawal = £100   **15** Savings = £40.25   Withdrawal = £45.32

EXERCISE 1.16

Complete the following problems and also write down whether the account is in credit or overdrawn.

1 £5.75+£8.26−£12   **2** £14.52+£36.92−£52.10

3 £14.32−£18.15+£3.27   **4** £32.50−£18.27−£13.25

5 £336.82−£52+£15.20   **6** £22.50−£3.25+£5−£26

7 Rachelle has £203.50 in the bank. She buys a new washing machine for £219.99. By how much is she overdrawn?

8 Mark's savings account has just £18.71 in it when he writes a cheque for £32.00. How much should he put back into the account to make it balance?

9 Calculate the amount Sadiah's account will be overdrawn from the statement below:

	Withdrawals	Balance
		£225.12
Cheque	£212.00	
Cheque	£132.17	

10 Jamie pays the following amounts into his account: £22.50, £101.20, £35.37, £40.36 and then pays the following bills: £48.13, £56.20, £80.40, £20.01. Find out the amount he is in credit or overdrawn.

Investigation I

Derek helps train a football team part-time for fifteen hours per week, but has a problem getting out of bed in the morning. He is paid £2.20 for every hour he works, and fined £1.70 for every hour he fails to turn up for work. How much is he paid for 15 hours' work?

How much is he paid for 10 hours' work?

What is the minimum number of hours he needs to work to get any payment at all?

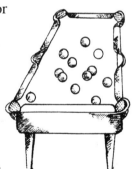

Investigation J

When playing games we always talk of the 'best of' 3, 5 or 7, etc. In snooker it is the best of 15 frames, 19 frames, etc. All these are odd numbers. Why do you think this is? Why shouldn't we have the best of 6, or 8?

Investigation K

When we play a game and the winner is to be the best out of 7, what is the minimum number of games which can be played before a winner is decided? For example, Alan has won 12 games and Fred 10, and they are playing for the best out of 35 games. What is *(a)* the minimum, and *(b)* the maximum number of games Alan must play to win? Try some other situations between your friends.

2. Money (2)

Simple fractions

(a) 'Here's 40p for your bus fares – give half to John.'
(b) 'Cheese is £1.80 a pound. How much will a quarter of a pound cost?'
(c) 'Commission on goods from a mail-order catalogue is $\frac{1}{8}$ of the payments. How much commission do I have if my total payments are £317.60?'

These are three typical problems in which you need to find a fraction of a sum of money. There are many ways of tackling these problems, but all of them depend on the realisation that the sum of money has to be divided by the **denominator** (bottom number) of the fraction, and multiplied by the **numerator** (top number) of the fraction. Of course, if the numerator is 1 then it's easy!

Now let us see if we can solve these problems.

(a) This is easy – John gets half of 40p which is 20p.
(b) We need $\frac{1}{4}$ of £1.80, so we divide £1.80 by 4, which gives £0.45 or 45p (remember that your calculator will not change £ into pence – you have to do it!).
(c) My commission will be worth £317.60 divided by 8, which is £39.70 (remember that 39.7 on your calculator means £39.70).

EXERCISE 2.1

Work out:

1 $\frac{1}{3}$ of £6.00   2 $\frac{1}{4}$ of £5.00   3 $\frac{1}{2}$ of £16.20

4 $\frac{1}{6}$ of £84   5 $\frac{1}{10}$ of £13.30   6 $\frac{1}{7}$ of £7.21

7 $\frac{1}{4}$ of £38.52   8 $\frac{1}{5}$ of £312.85   9 $\frac{1}{3}$ of £806.01

10 $\frac{1}{100}$ of £2.00   11 $\frac{1}{8}$ of £60   12 $\frac{1}{4}$ of £2.44

13 $\frac{1}{5}$ of 85p   14 $\frac{1}{9}$ of £14.31   15 $\frac{1}{20}$ of £80

16 $\frac{1}{8}$ of £4.00   17 $\frac{1}{3}$ of £3.99   18 $\frac{1}{2}$ of £19.98

19 $\frac{1}{6}$ of £7.26   20 $\frac{1}{1000}$ of £2000

Example 1

'Children pay $\frac{2}{3}$ of the adult rate of £2.40'. How much do you have to pay for a child?

To work out $\frac{2}{3}$ of £2.40, first find $\frac{1}{3}$ of £2.40, which is 80p (your calculator will give 0.8) and then multiply by 2, to give the answer of £1.60.

Example 2

Find $\frac{3}{4}$ of £6.00.

$$\frac{1}{4} \text{ of } £6.00 = \frac{£6.00}{4} = £1.50$$

So $\quad \frac{3}{4}$ of £6.00 $= 3 \times £1.50 = £4.50$

EXERCISE 2.2

Work out:

1 $\frac{2}{3}$ of £12.00	**2** $\frac{2}{5}$ of £20	**3** $\frac{3}{4}$ of £8.00
4 $\frac{5}{7}$ of £21.00	**5** $\frac{3}{4}$ of £5	**6** $\frac{4}{5}$ of £12.50
7 $\frac{3}{8}$ of £60.00	**8** $\frac{7}{10}$ of £28.40	**9** $\frac{5}{6}$ of £152.40
10 $\frac{4}{9}$ of £85.05	**11** $\frac{2}{7}$ of 56p	**12** $\frac{2}{3}$ of £120
13 $\frac{4}{7}$ of £35	**14** $\frac{3}{10}$ of £3.00	**15** $\frac{7}{8}$ of £4.00
16 $\frac{3}{4}$ of 72p	**17** $\frac{5}{8}$ of £100	**18** $\frac{2}{9}$ of £12.06
19 $\frac{9}{10}$ of 90p	**20** $\frac{5}{12}$ of £60	

EXERCISE 2.3

Now try these problems involving fractions.

1 'Jane, here's £30 between you and Mary. Give Mary a quarter of the money.' How much will Jane have to give Mary?

2 'This week your milk bill comes to £5.25, and you've had the same amount of milk on each of the seven days.' How much did one day's milk cost?

3 How much will it cost me to buy a metre of dress material, if I know that my friend bought three metres of the same material for £22.98?

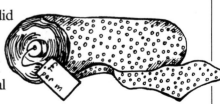

4 A group of six people win £83 472 on the pools. How much should each one receive?

5 A carton of twelve tins of fruit costs £5.88. How much does one tin of fruit cost?

6 'The grant will be $\frac{2}{5}$ of the total cost of building the sports pavilion.' If the total cost is £12 000, how much will the grant be?

7 If you wish to buy a car, you can borrow only $\frac{2}{3}$ of its price. How much can you borrow if you wish to buy a car priced at £6900?

8 How much will it cost me for $\frac{3}{4}$ of a metre of cable if one metre of cable costs £5.76?

9 About $\frac{4}{5}$ of the population of the United Kingdom live in England. If the population of the United Kingdom is 56 600 000, how many people live in England?

10 In 1976, milk cost $\frac{3}{8}$ of its 1986 price of 24p per pint. How much did a pint of milk cost in 1976?

The four rules

There are many occasions when we need to work out problems involving money, in which we need to add, subtract, multiply and divide. In some cases addition only is required, in others a combination of rules is needed in order to solve the problem.

Example 3

Mrs Harris has a balance of £1560 to pay on her new car. Calculate each instalment if she decides to pay the balance by (*a*) 12 equal monthly instalments, (*b*) 52 equal weekly instalments.

We need to divide in each of these cases.

(*a*) £1560÷12 = £130 per month
(*b*) £1560÷52 = £30 per week

EXERCISE **2.4**

Use methods you know to answer these questions.

1 I buy a bar of chocolate for 28p and some mints for 19p. How much is this altogether?

2 At the checkout my groceries come to £17.83. How much change will I get from four £5 notes?

3 To pass its MOT test my uncle's car needs a tyre and a handbrake cable. The tyre costs £22.74 and the handbrake cable costs £6.37. How much do these items cost together? What change will my uncle get from three £10 notes?

4 A car dealer sold three cars on one Saturday. The prices were £2150, £6895 and £1250. How much money did the dealer take on these three cars?

5 The new price of a bicycle is $\frac{1}{6}$ greater than the old price of £120. What is the new price?

6 You buy eight bars of chocolate at 27p each and eight packets of sweets at 34p each for your sister's birthday party. Your mother has given you a £5 note. Is this enough money? If so, how much change is there? If not, how much more do you need?

7 On holiday you buy seven postcards at 18p each and two at 23p each. You need to buy stamps at 20p to send the postcards. If you have £3 with which to buy postcards and stamps, how many stamps can you afford?

8 'Sale – 20% off – you pay only $\frac{4}{5}$ of the marked price.' How much would you pay for a pair of jeans with a marked price of £15?

9 For lunch you buy a beefburger, chips and beans, followed by apple pie and a yoghurt. How much change will you get from £1.50?

10 The return train fare to the football ground is 64p, and it costs £1.40 each to get into the ground. We buy one programme among the three of us, costing 50p. How much could we spend at the cafe at half-time if the three of us have £8.17 in total?

Todays menu
chips   30P
Fishfinger 15P
Sausage   15P
Beefburger 25P
Beans   12P
Egg   10P
cheesecake 23P
Apple pie   18P
Orange   12P
Yoghurt   12P
Biscuit   8P

11 My mum gave me £10 to buy a skirt and top when I went shopping with Aunt Emma. The skirt I liked cost £7.99 and the top was £4.49. How much did I have to borrow from Aunt Emma in order to buy the clothes?

12 Your friend desperately wants a particular record, costing £6.99. Her mother will give her 30p for every message that she takes, and also £1 for each night that she babysits. She has £3.23 saved already, and can babysit for two nights this week. How many messages will she need to take by Saturday if she wants to buy the record then?

13 Judith has a jar in which she saves money for her holidays. After a week she tips her jar out on to the table. There are four 20p coins, seven 10p coins, twelve 2p coins and five 1p coins. How much has she saved so far?

14 Ian and Judith each have a jar in which they save money for their holidays. The money is to be shared out equally among Ian, Judith and their three cousins. If Ian's jar contains £28.39 and Judith's contains £21.56, how much holiday money will each of the five children have to spend?

Investigation A

There are many ways of buying varying numbers of first and second class stamps if you have only a £1 coin. Investigate all the possible ways that there are. Lay out your findings systematically, stating the change you would be given in each case.

Extension

(a) How many ways give no change?

(b) Repeat the exercise, assuming that each stamp costs one penny more.

(c) How many more possibilities would there be if you had £2?

Investigation B

You can pay for a 24p comic in a number of ways (e.g. a 20p coin and two 2p coins, or two 10p and four 1p coins, etc). What is the smallest number of coins needed for each amount up to 99p? You may use any of the usual coins as many times as you need.

Extension

(a) If you have only *one* of each of the 1p, 2p, 5p, 10p, and 20p coins, which amounts up to 38p can you *not* make exactly?

(b) Repeat the investigation, but assume that you also have a 3p coin.

3. Money (3)

Percentages

A percentage is a fraction of 100. The percentage symbol % means out of 100: 1% is $\frac{1}{100}$. In just the same way 1p is £$\frac{1}{100}$, so 1% of £1 is the same as 1 penny, which makes percentage easier to work out in terms of money. If we want 5% of £1 it would be 5 pence.

Example 1

What is 5% of £3?

£3 is three times as much as £1, so 5% of £3 is 5×3 = 15p.

EXERCISE 3.1

1% of £1 is 1 penny
Find the following:

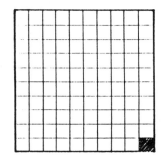

1 1% of £4	**2** 1% of £6	**3** 1% of £15	**4** 2% of £4
5 3% of £6	**6** 9% of £20	**7** 10% of £10	**8** 50% of £1
9 75% of £3	**10** 4% of £48	**11** 6% of £25	**12** 12% of £12

Many calculators have percentage buttons, but these can work in completely different ways. It is sometimes easier to remember how to work out percentages without using the percentage button.

Example 2

Find 4% of £5.

First of all we want to know what 1% of £5 is. As 1% means $\frac{1}{100}$, we divide £5 by 100:   5 ÷ 100
Having found 1% we now want 4%, that is 4 times as much, so we multiply by 4: × 4 = which gives us the answer.
So to summarise:

4% of £5: 5 ÷ 100 × 4 =
            to find 1%   for 4%

Use this method on your calculator to find 6% of £8. What buttons would you press to find 8% of £12?

If you understand how to use the percentage button well enough, use it.

EXERCISE 3.2

Find the following:

1 5% of £80	**2** 15% of £200	**3** 35% of £90	**4** 75% of £48
5 60% of £6	**6** 29% of £29	**7** 70% of £6.50	**8** 20% of £4.10
9 45% of £2.20	**10** $7\frac{1}{2}$% of £20	**11** $12\frac{1}{2}$% of £22.96	**12** $62\frac{1}{2}$% of £11.28

13 Matthew received 30% of the marks in a test out of 120. How many marks did he receive?

14 Julie receives 9% interest at the end of the year on the £85 in her savings account. How much interest is this?

15 A trainee salesman is paid $2\frac{1}{2}$% commission on his sales. How much will he get after a week's sales of £2024?

16 Exactly 14% of a crowd of 6350 at a hockey match were males. How many males attended?

17 A holiday company adds a fuel surcharge of 6% on the cost of all holidays. What would be the fuel surcharge on a holiday costing £217?

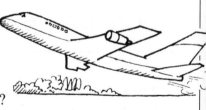

18 A loan company charges a fee representing $\frac{1}{2}$% of any loans for administrative costs. What would be the fee on a loan of £2500?

Approximation, to the nearest penny

Use your calculator to find 10% of £1.23 and write down your answer. You should have found the answer to be £0.123. What does this mean? What actual amount of money is it? Try and reach a conclusion with a friend.

Look at the figures more closely as we did in Chapter 1.

$$£0.123 = £0 \cdot \frac{1}{10} + \frac{2}{100} + 3 ?$$

The 3 represents $\frac{3}{1000}$. Is there anything wrong with this?
$\frac{1}{10}$ means a 10p coin, $\frac{2}{100}$ means two 1p coins.
$\frac{3}{1000}$ cannot be represented by money: it is less than one penny. We can see
this more easily by dividing up the pounds and the pence:

$$£0 \quad \cdot \quad 1 \quad 2 \quad 3$$
$$\text{pounds} \quad \text{pence} \uparrow \text{ fraction of one penny}$$

Frequently in money calculations we end up with fractions of one penny
which we do not use. We therefore **approximate** our answers to the
nearest penny, dropping the extra digits. This is also called rounding.

$$£0.12\underline{3} \quad \simeq \quad £0.12$$
$$£1.25\underline{41} \quad \simeq \quad £1.25 \quad (\simeq \text{means approximately equal to})$$

So far we have been rounding *down*. When the additional digits are half a
penny or more we usually round *up*.

$$£0.12\underline{6} \quad \simeq \quad £0.13 \leftarrow \text{add on one more penny}$$
$$£1.25\underline{55} \quad \simeq \quad £1.26$$

Whenever you have an answer which is too long you must approximate it
by rounding off to the nearest penny.

EXERCISE **3.3**

Approximate these amounts of money to make them sensible:

1 £1.124   **2** £3.345   **3** £5.017   **4** £0.106   **5** £4.994

6 £2.095   **7** £1.999   **8** £3.0132   **9** £4.1492   **10** £5.4555

11 £7.0123   **12** £0.6666   **13** £13.6709   **14** £4.15921   **15** £3.01243

Find the following percentages, rounding your answer when
necessary:

16 8% of £1.35   **17** 34% of £8.25   **18** 46% of £7.30

19 27% of £11.30   **20** 32% of £7.05   **21** 40% of £5.99

22 15% of £5.95   **23** 41% of £9.05   **24** $22\frac{1}{2}$% of £12.50

25 $9\frac{1}{2}$% of £45.30

Rounding to the nearest pound

Although rounding off to the nearest penny is the most common form of approximating money, it can be rounded in other ways.

To round to the nearest £, drop off all the digits after the decimal dot (the pence) and round up or down accordingly:

£1.23 ≃ £1
£2.67 ≃ £3

Note that there is no need to write in zeros for pence as there will be no pence after rounding off.

Investigation A

100 workers at a factory receive a pay increase of £2.30 each per week. A local newspaper article describes the pay award: 'Thomas Overalls have increased the women's wages by just over £2 per week – an increase of only £200 to the factory for the entire workforce.'

What have you to say about the accuracy of the article? Invent a similar situation yourself, and write a short newspaper article about it.

Rounding to the nearest £10

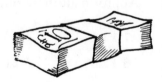

Round off to the nearest £, then further round off the £ units column.

£113.24 ≃ £11<u>3</u> ≃ £110 ← replace the units figure with a 0 after rounding

£25<u>6</u> ≃ £260

Why do you think we have to replace the units figures with a 0? What would the amount of money look like without the 0?

Rounding to the nearest £100

Round off to the nearest £10, then further round off the £ 'tens' column.

£113.24 ≃ £1<u>1</u>0 ≃ £100   ← replace the tens figure with a 0
after rounding

£256 ≃ £2<u>6</u>0 ≃ £300

EXERCISE 3.4

Round off these amounts of money to the nearest £:

1 £3.23   **2** £4.14   **3** £5.96   **4** £2.05   **5** £7.80

6 £14.35   **7** £101.30   **8** £23.77   **9** £49.65   **10** £99.82

Round off these amounts (*a*) to the nearest £10, (*b*) to the nearest £100.

11 £1215   **12** £2482   **13** £7889   **14** £8178   **15** £10 406

16 £6782.35   **17** £9433.92   **18** £10 099.13 **19** £14 407.50 **20** £20 010.30

Investigation B

Rounding can be used to deceive people. 'The profits of the company are £11 000 to the nearest £100.' The actual profits are £10 950. What is the difference between these two figures? Pick a figure which can be rounded down to the nearest £100. What is the difference between the two figures? Can you find the maximum possible difference in rounding down? Does it depend on the method of rounding, i.e. to the nearest £1, £10, £100, etc?

PROFIT BREAKDOWN   CONTRIBUTION TO PROFITS

EXERCISE 3.5

1 A summer fete produces a profit of £437.82. Write this to the nearest £.

2 The turnover of a company is £74 783. Round this to the nearest £10.

Summer Fete

3 Ten people are to share £247. How much will each receive, to the nearest £?

4 Door receipts at a concert are £1065.40, £3427.20 and £2274.40. What are the total receipts, written to the nearest £100?

5 A caretaker is to be paid £64.70. What is the maximum number of £10 notes he could have in his pay packet?

6 A case of twelve bottles of wine costs £105.48. How much is one bottle, to the nearest £?

7 Ronald has won £1 035 427.12. How much is this to the nearest £100?

8 A salesman claims 23p per mile as expenses on a journey. What would he claim for a journey of 195 miles, to the nearest £?

9 Mill Top School places a book order for 450 books costing £4.59 each, and 300 books costing £5.25 each. What is the total cost, to the nearest £10?

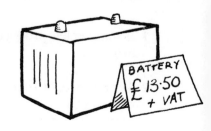

10 A house is valued at £80 430, and is to be advertised at a price written to the nearest £100. What will be the advertised price?

Value Added Tax (VAT)

Value Added Tax (VAT) is a tax which is added to most goods and services we buy. The actual rate of tax is determined by the Chancellor of the Exchequer, but the most common rate to date is 15%, or 15p for every £. By law the price of an article advertised should include VAT; if it does not then it should be clearly stated. If VAT is not included then you will have to add it on to find the actual price you will be charged.

Example 3

A car battery costs £13.50 plus VAT.

You need to find the VAT, that is, 15% of £13.50

$$£13.50 \div 100 \times 15 = £2.025 \approx £2.03 \text{ VAT}$$

So the full price to pay is £13.50 + £2.03 = £15.53

EXERCISE 3.6

Calculate the price of the following with the VAT added:

1 A pen marked £3 + VAT.　　**2** A pair of shoes costing £24 + VAT.

3 A football marked £11 + VAT.　　**4** A colour TV at £253 + VAT.

5 A digital watch at £9.65 + VAT　　**6** A camera costing £54.50 + VAT.

7 A lamp marked £54 + VAT.　　**8** A telephone bill for £39.80 + VAT.

9 A pair of jeans for £14.95 + VAT.　**10** A shirt marked £8.50 + VAT.

11 4 car tyres at £38 each + VAT.　**12** 5 records marked £8.50 each + VAT.

13 What is the total price for the van?

14 Find the total cost of having the car-phone installed.

15 Find the total price charged for each of the exhausts.

16 (*a*) Calculate the total price of the oven. (*b*) What is its normal price, including VAT?

MAESTRO 700 CITY VAN
'E' Reg - DELIVERY MILEAGE ONLY
£4895 + VAT

LARGEST DISPLAYS IN THE COUNTRY

LOOK — Quatra multi function built-under Elec oven + gas hob
RANGE "ALMOST ½ PRICE SALE
ONLY £199 + VAT
(½ NORMAL PRICE)

CAR PHONES
£695 + VAT
FULLY INSTALLED
NO HIDDEN EXTRAS
FREE FITTING + AERIAL

Top Quality
EXHAUSTS
COMPLETE SYSTEMS INC. FITTING

To fit Fiesta, Escort, Marina from £18.95 + VAT

To fit Maxi, Capri, Cavalier, Datsun Sunny, Renaults from £29.95 + VAT

To fit Granada, Alpine, Solara, Horizon from £48.95 + VAT

Investigation C

Derek works in a 'Cash & Carry'. He has a way of working out VAT in his head; he knows 15% is 10% + 5%, so he works out what 10% is, and then 5%, to make up the VAT. Work this out for a few different amounts, and try to find out why this might be a short cut for working it out in your head. Could you work out VAT in your head this way? Try a few and see.

Investigation D

Derek's mate Gerry finds working things out in his head difficult. He makes himself a ready-reckoner for VAT. Can you find out how to design a ready-reckoner so that you don't have to keep working VAT out?

EXERCISE 3.7

Copy and complete these bills:

INVOICE
Parts £39
Labour £120
VAT £23·85
TOTAL £182·85

1 *Mills Garage*

2 tyres 155 × 13	£45.60
1 valve	£0.90
2 wheel balances	£3.90
	£_____
+ VAT	£_____
	£_____

2 *The Copse Hotel*

Starters	£5.05
Main courses	£20.50
Sweets	£3.80
Drinks	£12.30
	£_____
+ VAT	£_____
	£_____

Make up similar bills for the following, for which the final cost must include VAT.

3 2 nights B & B at £9.50
2 dinners at £5.80
2 lunches at £3.70
Drinks: £3.50
all exclusive of VAT

4 Telephone bill:
System charge £13.95
345 units at 5p each
Misc. charges £5.80
all exclusive of VAT

5 Invoice:
3 crates soft drinks at £5.40 each
12 boxes crisps at £6.10 each
2 boxes choc. bars at £3.40 each
5 boxes of gum at £7.50 each
all prices exclusive of VAT

6 Invoice:
Paper: 8 reams @ £1.43 per ream
Wallet folders: 50 @ 72p each
Labels: 3 rolls @ £2.48 a roll
Paper: 2 rolls @ 90p per roll
Workbooks: 200 @ 5p each
all prices exclusive of VAT

Increase and decrease by a percentage

VAT is an example of an increase of an amount of money by a percentage. We can increase by any percentage.

The Meteor
INFLATION
9%

Example 4

Increase £24 by 45%.

We need to find 45% of £24:

£24 ÷ 100 × 45 = £10.80 (the increase)

So the increased amount is £24 + £10.80 = £34.80.

Similarly we can decrease by a percentage, which involves reducing the amount of money.

Example 5

Decrease £24 by 5%.

We need to find 5% of £24:

£24 ÷ 100 × 5 = £1.20 (the decrease)

So the decreased amount is £24 − £1.20 = £22.80.

EXERCISE 3.8

1 Increase £5.50 by 10%

2 Increase £12 by 25%

3 Decrease £20 by 35%

4 Decrease £40 by 30%

5 Increase £18 by 45%

6 Decrease £24 by 36%

7 Increase £60 by 32%

8 Increase £55 by 9%

9 Increase £30 by $12\frac{1}{2}$%

10 Decrease £40.50 by 30%

11 Increase £12.50 by 5%

12 Increase £10.20 by 35%

13 Decrease £4.80 by 75%

14 Decrease £19.80 by 45%

15 Decrease £15.50 by 6%

16 Increase £32.50 by 9%

17 Decrease £41.20 by 14%

18 Decrease £23.90 by 12%

19 Increase £90.70 by 4%

20 Decrease £14.99 by 25%

EXERCISE 3.9

1 A car manufacturer decides to increase all new car prices by $6\frac{1}{2}$%. What will be the new price of a model at £6550?

2 There is to be a reduction of 12% on a £450 hi-fi system. What will be the new price?

3 A pension of £42.50 per week is to be increased at the rate of inflation of 4%. What will be the new weekly pension?

4 Mark earns £115.60 a week, but has income tax deducted at a rate of 25%. How much does he actually get?

5 The cost of a £250 microwave cooker is expected to fall by 5% over the next year. What might it cost in one year's time?

6 A gas bill is £122.50. If prices are due to rise by 3.5% what would be the cost of the bill after the price rise if the same amount of gas is used?

7 Julie's wage of £83.25 is to be increased by $6\frac{1}{2}$%. What would be her new earnings?

8 A car tyre costing £80 is reduced by 10%, and then has VAT added at 15%. What is the actual cost of the tyre?

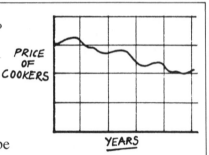

PRICE OF COOKERS

<u>YEARS</u>

Investigation E

£200 has been increased by 25% to give £250. Using your calculator, try to find a percentage decrease to reduce the £250 back to £200, that is by a reduction of £50. What does this tell you about percentage increase and decrease? Try some more different amounts of money and/or percentage.

Discount

Discount is a form of decrease in percentage. We receive discount when a price has been reduced, and it is worked out in the same way as a decrease.

Example 6

A second-hand car bought for cash with no trade-in receives a discount of 10%. The car would normally cost £850. Find the discount price.

$£850 \div 100 \times 10 = £85$ discount

So the discounted price is £850 − £85 = £765

Example 7

A computer is advertised at £395 exclusive of VAT, but receives a discount of 5% in a sale. Find the actual price paid.
Note: in this case we *cannot* say there is an overall increase of 10% (15% VAT − 5% discount). We have to work the problem out in two separate stages. VAT is charged on the final bill.

Discount: 5% of £395 = £395 ÷ 100 × 5 = £19.75 so price is £375.25
VAT: 15% of £375.25 = £375.25 ÷ 100 × 15 = £56.2875 ≈ £56.29
So actual price is £375.25 + £56.29 = £431.54

EXERCISE 3.10

1 A discount of 12% is given when a gas fire is traded in for a new one costing £65. What is the actual price to be paid?

2 Find the new price if a £216 computer receives a discount of $12\frac{1}{2}$%.

3 A builder offers a discount of $7\frac{1}{2}$% if he is paid in cash. What saving would be made on a £750 bill?

4 A book valued at £18.50 is given a discount of 16% through a book club. What will a book club member be charged for the book?

5 In a sale a pair of shoes at £8.75 is given a discount of 20%. What is the sale price?

6 A man is charged £1.80 for each parcel he has delivered, but is given a discount of 20% on deliveries of 50 or more. What saving will he make if he asks for 50 parcels to be delivered together?

7 A carpet costing £250 receives a discount of 30%, but is exclusive of VAT at 15%. Find the actual price to be paid.

8 A £48 000 house found to have dry rot is sold at a discount of 14%. What is the new price?

9 A car originally valued at £6200 has its price increased by 5% by the manufacturers, but is then sold at a discount of 4% off its new price. How much would it cost to buy the car?

10 A shop can claim a discount of 20% on a £320 video recorder, but it is then charged 12% for a credit-card sale. What is the net sale price of the video recorder?

Investigation F

Check the percentage decreases on the advertisement. What do you find? How do you explain it?

Investigation G

In a problem, discount is calculated first, then VAT. Try various combinations of discount and VAT to change the price. What percentage discount would balance the 15% VAT to make the price remain the same? Try different discounts to see if you can find the answer.

Profit and loss

Many businesses are concerned with buying and selling: they buy goods or articles, or produce them out of materials, and then they sell them hoping to make a profit in the process. The money paid out in buying materials and/or manufacture is called the cost or expenditure, while the money collected in sales is called revenue.

Example 8

Jenny buys a knitting machine for £450, and a stock of wool for £120. She produces 90 jumpers, and sells them all for £6.50 each.

$$\begin{aligned} \text{Total revenue} &= 90 \times £6.50 = £585 \\ \text{Total costs} &= £450 + £120 = \underline{£570} \\ \text{Profit} &= £15 \end{aligned}$$

If instead she had sold the jumpers for £6 each the situation would have been different:

$$\begin{aligned} \text{Total revenue} &= 90 \times £6.00 = £540 \\ \text{Total costs} &= £450 + £120 = \underline{£570} \\ \text{Profit} &= -£30 = £30 \text{ loss} \end{aligned}$$

The minus sign indicates money owing, which is a *loss* rather than a *profit*.

EXERCISE 3.11

For each problem calculate the result of the enterprise, clearly stating whether your answer is a profit or a loss.

1   Mr Hardcastle buys a box of 20 ornaments for £15, and sells them off at 80p each.

2 Jack buys four antique chairs for £18.50 each, and a table for £40, and sells them to an antique shop as a dining set for £120.

3 Martin purchases three old cars for a total of £480, and re-sells them for £160, £180 and £165.

4 Gillian's enterprise scheme entails buying £5 of ingredients for baking plus £3 for tins. She bakes 115 fairy cakes and sells them all for 7p each in the tuck shop.

5 Remi makes ornamental candles which he can sell at £1.50 each. For every 50 candles he estimates it costs him £55 for the wax, £5.50 for the wicks, and a further £15 for electricity to run the wax heaters.

6 Julie breeds maggots for fishermen. She sells them at £3.50 for every 100. She estimates it costs her £33 to produce 1000 maggots.

7 Thu runs a mobile disco. His monthly costs include an £80 payment on the equipment, £35 for records, and £5 in maintenance. For each gig he charges £65, of which £10 is transport costs. He is booked to do five gigs on average each month.

8 Barry makes fibreglass garden pools for £14 each. The materials he needs for each 20 he makes are: fibreglass sheets £140, resin £130, brushes £15, pots, etc. £8.

EXERCISE 3.12

Find the total amount of money being paid in, the total amount being paid out, and hence the profit or loss made for each of the problems.

1

Date		Received	Paid out
Oct 1	Balance	400.15	
Oct 5	New stock		233.50
Oct 9	Sales	50.47	
Oct 15	Advertising		80.80
Oct 16	Sales	40.89	
Oct 23	Sales	43.70	

2

	Cash in	Cash out
Feb 8	39.40	
Feb 9		40.10
Feb 16	5.27	
Feb 18	4.86	9.20
Feb 20	2.80	3.50

3

		Credit	Debit
Apr 1	Balance	142.66	
Apr 13	Chq 569147		63.45
Apr 14	Hill Top a/c	126.55	
Apr 16	Chq 569148		110.42
May 4	Chq 569149		60.99
May 5	S.Waters a/c	72.04	
May 9	Chq 569150		43.78

4

	SALES			EXPENSES		
	Parts	Showroom	Garage	Parts	Showroom	Garage
Nov 9	35.42	6542.14	692.40	240.00	—	135.20
Nov 10	48.09	—	405.99	—	—	101.37
Nov 11	23.10	10493.87	517.83	8.40	26504.33	120.47
Nov 12	60.89	6542.14	803.01	120.09	—	201.49
Nov 13	40.03	8402.21	712.00	—	—	201.01

Hire purchase

Hire Purchase (HP) involves putting down a **deposit** and paying the remainder of the price by weekly or monthly instalments. Due to the interest charged on the money being paid off it would be cheaper to pay cash for the article being bought, but HP is an alternative way of purchasing if you don't want to pay the full price at once. Can you think of any other alternative ways of finding the money?

Example 9

Remi wants to buy a £350 television set on HP.
The HP agreement demands a 10% deposit, plus 20 weekly payments of
£18.75. (*a*) What is the total HP price?

$$10\% \text{ of } £350 = £350 \div 100 \times 10 = £35$$

$$\text{Weekly payments} = 20 \times £18.75 = \underline{£375}$$
$$£410$$

(*b*) How much more does it cost to buy through HP than paying cash?

$$£410 - £350 = £60$$

EXERCISE 3.13

Find the total price payable under an HP agreement for the
following:

	Article	Cash price	Deposit	Payments
1	Suit	£150	$\frac{1}{5}$	10 × £14.08
2	Washing machine	£240	20%	12 × £17.60
3	Furniture	£400	£70	20 × £18.56
4	Stereo	£260	$\frac{1}{4}$	12 × £19.85
5	Gas cooker	£290	£75	24 × £10.95
6	Motor cycle	£550	30%	36 × £12.62
7	Camera	£120	£25	20 × £6.18
8	Caravan	£3600	20%	36 × £99.20
9	Freezer	£220	£45	12 × £16.35
10	Electric organ	£580	$\frac{1}{5}$	20 × £26.68

For each of the following problems find (*a*) the total amount payable
under HP, (*b*) the amount to be saved if cash is paid.

11	Bicycle	£130	£20	12 × £12
12	Greenhouse	£850	30%	24 × £30.05
13	Radio	£57	$\frac{1}{3}$	10 × £4.14
14	Television	£320	£50	12 × £24.75
15	Vacuum cleaner	£94	$\frac{1}{8}$	20 × £4.55

EXERCISE 3.14

1 Giles buys a central heating system for £2800. He needs 15% as a deposit, and will pay the remainder as monthly instalments of £50.98 over a five year period. Find how much he will have to pay.

2 Waheed's computer system is paid for by 24 instalments of £30.50, following a deposit of £50. The cash price is £650. How much will he save by paying cash?

3 A carpet is advertised with a cash price of £520, or with terms available on HP of a 10% deposit, plus monthly instalments of £15.86 over three years. Calculate the difference between the cash price and the total amount payable on HP.

4 Trisha is buying a piano for £5025. She needs a deposit of £400, and will repay the remainder as 50 payments of £109.15. How much will she save by paying cash?

5 Jason has seen a car for a cash price of £6870, but can only afford a deposit of £800. He agrees to pay the remainder as monthly instalments of £131.72 over a five year period. Find the total amount Jason will have to pay.

6 Calculate the HP cost of buying the car, and the difference between the HP cost and the cash price.

JUST **£3,050** ON THE ROAD

20% Deposit

Monthly Repayments
(48 Months) £69.13

7 The HP on the two articles requires no deposit, but is payable over 30 months.

   (a) Find the difference between the HP price and the cash price for the midi system.

   (b) How much more is the sale price of the CD system compared to the midi system?

High-Power Midi Hi-Fi System
Massive 50 watts RMS per channel power output
High speed dubbing Twin cassette decks 15 pre-sets
DOLBY B Noise Reduction   Was £349.99
Digital tuner Model Compact 48   SALE PRICE **£329.99**

NO DEPOSIT INSTANT CREDIT ONLY £14 MONTHLY

Twin-Decks CD System SAVE £100
Superbly sophisticated remote control system
with CD player Model: Compact 78 Was £699.99   SALE PRICE £599.99

4. Money (4)

Proportion

Example 1

I need six oranges for our football team at half-time. The greengrocer is selling them at three for 32p. How much will it cost for six?

It is easy to see that it will cost $2 \times 32p = 64p$, as we are buying three oranges 'twice'.

Example 2

If it takes twelve balls of wool to knit a dress, and four balls of wool cost £2.98, how much will the wool cost for the dress?

Here we need to buy four balls of wool three times (as $12 \div 4 = 3$): so it will cost $3 \times £2.98 = £8.94$.

EXERCISE 4.1

1. How much will 15 pencils cost, if five cost 39p?

2. A pack of two notebooks costs 99p. How much will I have to pay for eight notebooks?

3. A set of ten cards costs 73p. If I buy fifty cards, how much will it cost me?

4. Six eggs cost 53p. To cater for a party, I need three dozen (36) eggs. How much will these be?

5. A packet of nine marbles costs 34p. How much will it cost me to buy 63 marbles?

6. If 250 ml of white spirit costs 89p, how much will one litre (1000 ml) cost?

7. Three metres of twin-core cable cost £1.25. How much will it cost me to buy 24 metres of twin-core cable?

8. Four cans of lemonade cost 87p. How much do sixteen cans cost?

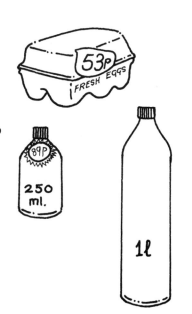

9 150 cm of chain costs £2.14. I need 750 cm of chain. How much will this cost?

10 A box of 6 bobbins of thread costs £3.49. How much will 30 bobbins cost?

Example 3

My bedroom walls require seven rolls of wallpaper. Yesterday, my next-door-neighbour bought four rolls of the same type of wallpaper, and it cost her £22.00. How much will it cost me for the seven rolls I need?

In this case it is less easy to compare the two quantities of wallpaper. If we find out, however, the cost of **one** roll of wallpaper, we can easily work out the cost of seven (or any number) of rolls, by multiplying.

As four rolls cost £22.00, one roll will cost $\frac{1}{4}$ of £22.00 which is £5.50 (£22.00 ÷ 4 = £5.50 — remember that your calculator will give 5.5 which means £5.50). So seven rolls will cost 7 × £5.50 = £37.50.

EXERCISE 4.2

1 Five calculators cost £30. How much would three cost?

2 Eight copies of a piece of music cost £20. How much would it be for five copies?

3 If three tubs of cream cheese cost £1.08, how much would four tubs cost?

4 A carton of 12 tins of fruit costs £4.80. If I want to buy seven tins, how much will I have to pay?

5 A group of four friends went to a concert. The four tickets cost £17 altogether. Another group of seven people also went to the concert. How much was the total cost of the tickets for this group?

6 There is a special offer price of £7.00 for a pack of 20 light bulbs. I buy a pack. How much should I charge a neighbour who has said that he wants to buy nine from me?

7 A local village shop is selling trays of 30 free-range eggs for
£1.20 a tray. If a woman buys a tray, how much should she
charge her neighbour for a dozen (12) eggs?

8 My last newspaper bill was for five weeks and came to £8.75.
My next bill will be for seven weeks. How much will it be?

9 A set of eleven cups for a cricket team costs £13.75. How much
would it cost for similar cups for a tennis team of five players?

10 I buy 12 square metres of bedroom carpet for £154. If I intend to
buy the same carpet for another bedroom whose floor area is
15 square metres, how much will it cost me?

Example 4

A 5 kg bag of plaster costs £7.80. How much would a 1.5 kg bag cost?

Here we can find the cost of one kilogram, as before, by dividing £7.80 by
5 to give £1.56. Now multiply £1.56 by 1.5 to give the answer of £2.34.

In some problems the figures do not work out exactly. In these cases it is
normal to give the answer correct to the nearest penny – check back to
Chapter 3 if you need to remind yourself of what to do.

EXERCISE 4.3

1 A large piece of pork weighing 3 kg costs £4.83. How much
would it cost to buy a smaller piece of pork, weighing 1.8 kg?

2 I estimate that I will need 3.6 litres of paint for my fence. How
much will it cost me if a two-litre tin costs £4.69?

3 The instructions for making 24 bottles of beer tell you to add
1.5 g of priming sugar. How much priming sugar should be
added if you have a kit for making 40 bottles?

4 One metre of cable costs £1.26. I need 24 inches of cable for a
model I am making. Knowing that one metre = 39.37 inches,
how much will the cable for my model cost me?

5 A palette of 550 bricks would cost £33. If I need to buy 200, how much should I have to pay?

6 If your friend has bought a pack of 50 envelopes for £1.19, how much should you pay her if you want to buy 18 envelopes from her?

7 A tin of 326 g of sweetcorn costs 47p. How much would a portion weighing 100 g cost?

8 A length of skirting board costs £2.49, and is 1.83 m long. I estimate that I need 9.7 m of skirting board for a room. How much will it cost me?

9 A 5 litre bottle of white spirit cost me £2.99. My neighbour says that he will buy 2.25 litres from me. How much should I charge him?

10 The coalman delivered 30 kg of coal to my next-door-neighbour which cost him £12.66. How much should I pay my neighbour, if I want to buy 7 kg of coal from him?

Investigation A

My house needs rewiring. I am likely to need various lengths of three types of electric cable. The cost of one metre of each cable is stated below. Make up a table which will enable me to read off the cost of any length, to the nearest 10 cm, of any of the three cables, up to 10 metres in length.

> 5 amp: 39p per metre;
> 30 amp: 69p per metre;
> 40 amp: £1.99 per metre.

Ratio

If I have £6.00 to spend, Tasneem has £9.00 and Tom has £18.00, we can compare our spending money by saying that Tom can buy twice as much as Tasneem, or three times as much as I can.

The way that we compare is to write the sums of money as a ratio – that is with a :

$$£6.00:£18.00 = 6:18$$
$$= 1:3 \quad \text{(dividing each number by 6)}$$
Also $\quad £9.00:£18.00 = 9:18$
$$= 1:2 \quad \text{(dividing each number by 9)}$$

When we want to compare my £6.00 with the £9.00 Tasneem has, we can say

$$£6.00:£9.00 = 6:9$$

This time, dividing by the smaller number (6) will not give a whole number on both sides – it will give 1:1.5, which later we will find can be a very useful method for comparing different ratios.

For the present, we will try to obtain whole numbers on each side of the ratio. We can write:

$$6:9 = 2:3 \quad \text{(dividing each number by 3)}$$

Example 5

Write the ratio 20:32 as simply as possible.

We could divide both numbers by 2, which would give 10:16. Then we could divide by 2 again, giving 5:8.

Had we divided by 4 in the first place, we would have reached the answer of 5:8 in one step.

It is quicker to try to find the highest number that will divide exactly into both numbers, i.e. the highest factor of both numbers.

The order in which the numbers of a ratio are written down is vital. A ratio of 4:7 is not the same as a ratio of 7:4. Make sure that you write them down in the correct order.

EXERCISE 4.4

Write each of these ratios as simply as you can.

1 4:8	**2** 2:10	**3** 5:15	**4** 14:7
5 9:3	**6** 12:16	**7** 10:60	**8** 30:18
9 6:21	**10** 20:30		

Example 6

Simplify £9:£24.

We can divide each amount by 3, and as in the initial example we can ignore the £ signs. This is because a ratio is a comparison of numbers of the *same* thing (in this case, £).

So   £9:£24 = 3:8

We would have arrived at the same ratio if we had simplified 9p:24p or £90:£240. We need to be very careful, however, if we have to simplify a ratio of 90p:£240, because we are comparing *different* things, £ and pence.

Example 7

Simplify £2:50p.

Here we *cannot* say 2:50 = 1:25, because we are not comparing the same things.
Working in pence

$$£2:50p = 200p:50p$$
$$= 4:1$$

Working in £
$$£2:50p = £2:£0.50$$
$$= £4:£1$$
$$= 4:1$$

EXERCISE 4.5

Simplify these money ratios:

1 30p:60p   **2** 18p:54p   **3** £4:£10

4 £12.50:£2.50   **5** £12:£18   **6** 80p:50p

7 £1.50:75p   **8** 20p:£2   **9** £16:£36

10 £8:£1.60

EXERCISE 4.6

1 I have £5 and my sister has £15. What is the ratio of my money to my sister's?

2 For every £1 that I earn, my father says that he will give me an extra payment of 20p. What is the ratio of my earnings to my father's extra payment?

3 The adult train fare to town is £2.40 and the fare for a child is £1.50. Write the ratio of the adult fare to the child's fare as simply as you can.

4 A job in a supermarket is advertised at £2.10 per hour from Monday to Friday and £3.50 per hour on Saturdays. Work out the ratio of weekday pay to Saturday pay as simply as possible.

5 My watch cost £9 and my friend Lien's cost £12. Simplify the ratio of the cost of my watch to the cost of Lien's.

Best buys

As stated before Example 5, we could write the ratio 6:9 as 1:1.5, by dividing both numbers by 6. We can do this for any ratio (that is, dividing both numbers by one of them); this will enable us to compare ratios more easily. In particular, we can use this idea to find out 'best buys'.

Example 8

A 300 g jar of coffee costs £4.85 and a 200 g jar costs £3.24. Which jar is the better buy?

By finding out how much a hundred grams of coffee costs in each case, we can compare their values more easily.

300 g jar: As 300 g cost £4.85,

then 100 g costs $\dfrac{£4.85}{3} = £1.616\,666\,6\ldots$

200 g jar: As 200 g cost £3.24,

then 100 g costs $\dfrac{£3.24}{2} = £1.62$.

Although the 300 g jar is just the better buy, there is little difference in value.

Example 9

Which is the better buy, a 3 kg bag of potatoes costing 49p or a 5 kg bag costing 70p?

Working out how much 1 kg of potatoes from each bag would cost:

3 kg bag: As 3 kg cost 49p,

then 1 kg costs $\dfrac{49\text{p}}{3} = 16.333\ldots$ p or 16.3p per kg (approx.)

5 kg bag: As 5 kg cost 70p,

then 1 kg costs $\dfrac{70\text{p}}{5} = 14\text{p}$

So the 5 kg bag is the better buy at 14p per kg instead of 16.3p per kg.

EXERCISE 4.7

Work out which is the better buy in each case:

1 2 kg of sugar for 54p or 5 kg of sugar for £1.25.

2 A 75 cl bottle of wine for £2.49 or a 2 litre (200 cl) bottle of the same wine for £5.98.

3 3 rolls of wallpaper for £10.50 or 10 rolls for £36.00.

4 6 golf balls for £8.49 or 10 golf balls for £14.99.

5 4.5 m of material for £7.49 or 7.2 m of material for £11.98.

6 A 326 g jar of marmalade for 66 p or a 1 kg tub for £1.99.

7 6.7 ounces of cheese for 91 p or 11.3 ounces for £1.61.

8 A 454 g tin of beans for 35 p or a 'family size' 1 kg tin for 79 p.

9 A pack of three blank video tapes for £7.99 or a pack of five for £12.99.

10 4 kg of potatoes for 59 p or a 25 kg sack for £3.50.

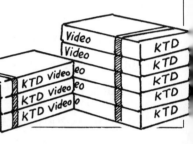

Investigation B

Choose a grocery product that comes in at least three sizes (tins of beans, coffee, potatoes, tea-bags, etc.). Investigate which size is the best buy in at least two shops or supermarkets.

Investigation C

Any ratio (e.g. 7:3) can be written as $\frac{7}{3}$:1 or 2.333 … :1.

Investigate which numbers as *denominators* (the '3' in this example) will **always** give a finite answer, whatever the numerator may be. (In this case the '3' does *not* give a finite answer, because 2.333 … does not terminate, that is, it goes on for ever.)

Formulas

A formula is a rule for working out some value, using numbers and letters instead of words.

Example 10

A formula for the cost of fencing is:

$$\text{cost} = £5.00 + (n \times £1.20)$$

where n stands for the number of metres of fencing required. Find the cost of 4 m of fencing.

$$Cost = £5.00 + (4 \times £1.20)$$
$$= £5.00 + £4.80$$
$$= £9.80$$

EXERCISE 4.8

Evaluate (work out the value of) these formulas:

1 $L = £62 + (k \times £6)$, when $k = 8$.

2 $P = £2.75 + (r \times £1.50)$, when $r = 10$.

3 $C = (p \times £48) + £73$, when $p = 9$.

4 $X = (h \times £2.64) - £20$, when $h = 15$.

5 $G = (a \times 65p) + (b \times 30p)$, when $a = 3$ and $b = 5$.

6 $T = £120 + (f \times £2) + (g \times £5)$, when $f = 27$ and $g = 20$.

7 $K = s \times (£4.50 + t)$, when $s = 30$ and $t = £5.50$.

8 $A = (d + 3) \times £79.37$, when $d = 8$.

9 $Z = £300 - (w \times £5.50)$, when $w = 24$.

10 $V = (k + 16) \times £1.61$, when $k = 14$.

Example 11

For babysitting I am paid £1.20 an hour.
(a) How much will I be paid if I babysit for 3 hours on Monday, $4\frac{1}{2}$ hours on Thursday, and 4 hours on Saturday?
(b) Write down a formula (rule) for working out how much I would be paid for babysitting for n hours.

(a) 3 hours on Monday will be worth $3 \times £1.20 = £3.60$.
$4\frac{1}{2}$ hours on Thursday will be worth $4\frac{1}{2} \times £1.20 = £5.40$
(remember $4\frac{1}{2} = 4.5$, and 5.4 is £5.40).
4 hours on Saturday will be worth $4 \times £1.20 = £4.80$.
Altogether I will be paid $£3.60 + £5.40 + £4.80 = £13.80$.
(b) It is clear from the examples above that the number of hours (n) has to be multiplied by the hourly rate of pay (£1.20) to give the total pay for n hours. The formula, or rule, is

$$pay = n \times £1.20$$

Example 12

A box holds 12 cans of lemonade. How many cans are there in
(*a*) 3 boxes, (*b*) 8 boxes, (*c*) 20 boxes, (*d*) *y* boxes?

(*a*) 3 boxes will hold $3 \times 12 = 36$ cans.
(*b*) 8 boxes will hold $8 \times 12 = 96$ cans.
(*c*) 20 boxes will hold $20 \times 12 = 240$ cans.
(*d*) *y* boxes will hold $y \times 12$ cans.

EXERCISE 4.9

1 How much will I be paid for 5 hours of stacking shelves if the rate of pay is £1.35 an hour? Write down a formula for calculating my pay for working *x* hours.

2 'Secretary required at £3.58 per hour.' How much does this job pay for a $7\frac{1}{2}$ hour day? Write down a formula for the secretary's earnings if she works for *t* hours.

3 An insurance agent is paid £537.50 on the 28th of each month. What is his annual salary? Write down a formula for calculating the salary of an insurance agent who works for *m* months.

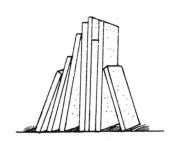

4 A formula for the cost of *d* square metres of chipboard is stated as $d \times £0.63$. Work out the cost of (*a*) 2 square metres, (*b*) 12 square metres, (*c*) 20 square metres.

5 The formula for average speed is $\frac{d}{t}$, where *d* is the total distance travelled and *t* is the total time taken. Work out the average speed when (*a*) $d = 20$ km and $t = 4$ hours, (*b*) $d = 240$ miles and $t = 6$ hours, (*c*) $d = 74$ m and $t = 10$ seconds.

6 My aunt has given me a box of 360 old buttons. (*a*) If I share them equally among three of us, how many will each of us have? (*b*) If instead, I share them among four of us, how many will each of us have? (*c*) Again, if there are now six of us, how much is each share?
   If there are *p* people, write down a formula for working out the number of buttons that each person would have.

Example 13

Mrs Black's electricity bill is made up of two parts: a quarterly charge of £8.00, and a charge of 5.25p for each unit of electricity used during the charging period (3 months). Work out a formula for calculating Mrs Black's bill.

In words,

   bill = quarterly charge + (number of units used) × (cost of one unit)

We can shorten this by using letters instead of words:

   $B =$ Q $+$ n $×$ c
or $B = Q + (n × c)$

The brackets indicate that the value of $n × c$ is added on to the value of Q. Without the brackets the formula would be $B = Q + n × c$, and you might add Q and n first, and then multiply by c. This would be wrong.

   So a formula for Mrs Black's bill is

   $B = £8.00 + (n × 5.25p)$

If Mrs Black used 1000 units, her bill would be

   $£8.00 + (1000 × 5.25p)$

You need to be careful when using your calculator because one figure is in pounds (Q) while the other ($n × c$) is in pence (see Chapter 3).

   B = £8.00 + 5250p
     = £8.00 + £52.50   (dividing the pence by 100 to give £)
     = £60.50

(We could have said that £8.00 = 800p, and arrived at the answer of 6050p, but because an electricity bill is much more likely to be stated in pounds, we may as well work in pounds.)

EXERCISE 4.10

1   What would Mrs Black's bill have been if she had used 1200 units?

2   Her neighbour, Mrs Sahir, used 1336 units. What was her bill?

3   Mr Robinson used 2317 units. Work out his bill, correct to the nearest penny.

4   For the next three months the quarterly charge increased to £9.50, and the cost per unit went up to 5.48p.

   (*a*)   How much was Mrs Black's bill for these three months, if she used 1513 units? State your answer correct to the nearest penny.

   (*b*)   Write down a formula for the bill for these three months if n units are used.

Example 14

To hire a car, it costs £25 basic charge plus 28p for every kilometre driven in it.

(*a*) Write down a formula for the hire cost if you travel a distance of k kilometres.

(*b*) Use the formula to calculate the hire cost of a journey of 200 km.

(*a*) In words,

hire cost = basic charge + (number of kilometres × 28p)

$$= £25 + (k × 28p)$$

(*b*) For 200 km,

$$
\begin{aligned}
\text{cost} \quad &= £25 + (200 × 28p)\\
&= £25 + (5600p)\\
&= £25 + £56 \qquad \text{(dividing by 100 to change pence into}\\
&\qquad\qquad\qquad\qquad\text{pounds)}\\
&= £81
\end{aligned}
$$

EXERCISE 4.11

1 I hired the car in Example 14 for the weekend. Altogether I travelled 341 km. How much did it cost me to hire the car?

2 The next person to hire the same car drove it for 89 km. How much did it cost to hire?

3 For a smaller car, the basic charge is £23.50, and the charge per kilometre is 24p.

    (*a*) Write down a formula for the cost of hiring this car for a journey of x kilometres.

    (*b*) If I had used this car instead of the car I used in question 1, how much would I have saved on the trip of 341 km?

4 I can afford £50 to hire a car.

    (*a*) Can I use the first car for a trip of 100 km?

    (*b*) Will I be able to use the second, smaller car?

Large numbers, rounding

Example 15

There were 33 272 people at a football match, who each paid £5 for a ticket. How much money did the club receive from the sale of these tickets?

The club received £5 × 33 272 = £166 360.

These are large numbers, and we can often get a clearer picture by making approximations, stating only the most important (significant) figures.

In Example 15, we could have said that there were over 33 000 people, and the club took over £166 000.

Example 16

Write these numbers correct to the nearest thousand: (*a*) 5180 (*b*) 42 761 (*c*) 2 485 830.

If the last three figures are less than 500, then they are replaced by three 0's. So in (*a*), 5180 is written as 5000, to the nearest thousand.

If the last three figures are 500 or over, then the 'thousands' figure goes up by one, and the last three figures are again 0's. So in (*b*), 42 761 becomes 43000, and in (*c*) 2 485 830 becomes 2 486000, to the nearest thousand.

EXERCISE 4.12

1 On the first Saturday in October the total attendance at soccer matches was 386 523. On the next three Saturdays the attendances were 412 701, 318 080 and 369 927. How many attended, in total, during October? Give your answer correct to the nearest thousand.

2 A 'jackpot' payout of £1 275 363 was shared amongst three people. How much did each receive, correct to the nearest thousand pounds?

3 If the number of unemployed people was 3 216 388, and $\frac{4}{5}$ of these were male, how many men were unemployed? Give your answer correct to the nearest 10 000.

4 During 1986, North Sea oil production was 2 560 000 barrels per day. If the average price of a barrel of oil during 1986 was $22.3, estimate the income from selling these barrels. State your answer correct to the nearest million dollars.

5 In 1986 there were 201 396 homes built in Britain, $\frac{4}{5}$ of which were for the private sector. How many private homes were built, correct to the nearest thousand?

6 A window measures 457 mm by 958 mm. Work out the area of the window, correct to the nearest 100 square millimetres.

7 My uncle has just bought a new house. He has taken out a mortgage with a building society, which means that he has to pay £352.84 on the 14th day of each month for a period of 25 years. How much will he pay in total, correct to the nearest thousand pounds?

8 It is one mile from one end of a village to the other. On the day of the Summer Fayre it is intended to lay a 'mile of pennies' through the village. Assuming that the length of the line of pennies is 1500 metres and that a penny is 2 cm in diameter, estimate the value of the 'mile of pennies'.

9 It is about a quarter of a million miles from the earth to the moon. Write this distance (*a*) as a six-figure number, (*b*) in kilometres, to the nearest 100 km, if 8 kilometres = 5 miles.

10 'Ninety thousand fans watched the Cup Final, and fifty-eight thousand watched the replay.'

(*a*) Write in figures the total attendance at both matches.

(*b*) If the average price of a ticket was £7.50, how much money was taken in total? Give your answer correct to the nearest £100.

Revision exercises: Chapters 1-4

1   Write down the common factors of 6 and 12.

2   Write down the common factors of 8 and 30.

3   List the first six common multiples of 5 and 15.

4   List the first six common multiples of 4 and 6.

5   Write down the factors of 48 which are odd numbers.

6   Finds all the prime numbers between 30 and 50.

7   Write down all the prime numbers between 10 and 30.

8   List all the multiples of 3 which are even numbers and which are less than 50.

9   Write down all the multiples of 7 less than 70.

10   Find all the factors of 24.

11   A box of chocolates is priced in four shops as £2.17, £2.23, £2.09 and £2.19. List these prices in order, starting with the cheapest.

12   Place the following amounts in order of size, smallest first: £0.99, £0.09, £1.09 and £9.99.

13   How many pence are there in (*a*) £1.27 (*b*) £0.32 (*c*) £14.05?

14   How many pence are there in (*a*) £2.31 (*b*) £12.50 (*c*) £0.08?

15   Write in pounds and pence (*a*) 134p (*b*) 1000p (*c*) 19p.

16   Write in pounds and pence (*a*) 750p (*b*) 232p (*c*) 98p.

17   Write as a decimal amount of money (*a*) £$1\frac{17}{100}$ (*b*) £$6\frac{3}{10}$.

18   Write as a decimal amount of money (*a*) £$\frac{29}{100}$ (*b*) £$4\frac{7}{10}$.

19 Write as a decimal amount of money (*a*) £$\frac{9}{100}$ (*b*) £$8\frac{21}{100}$.

20 I have £1.60, Ben has half as much as I have, and Bill has half as much as Ben. How much do the three of us have altogether?

In questions 21–50, work out the answers to the sums.

21 £24.57 + £5.74 − £7.09   **22** £14.42 + £5.99

23 £2.20 + £5.30   **24** £16.10 − £2.90

25 £5 + £2.30 + 23p   **26** £6.32 − £2.03 + £9

27 £9.32 − £5.12   **28** £8 − 80p

29 £5.35 + £2.17 + £3.12   **30** £1.11 + 39p − £1.01

31 £10.00 − £5.37 − £1.88   **32** £12.98 + £6.49 − £7.84

33 £69.50 − £4.75 − £9.99   **34** £40 − £15.98 + 68p

35 £6.66 + £0.66 + £6.06   **36** £98 − £76.54 + £32.10

37 £1.35 × 10   **38** £4.30 ÷ 10

39 £3.27 × 5   **40** £2.75 × 100

41 £11.64 ÷ 12   **42** £5.45 ÷ 5

43 £1.31 × 9   **44** £3.40 × 100

45 £30 ÷ 100   **46** £14.28 × 7

47 £19.98 × 4   **48** £4 ÷ 100

49 16p × 10   **50** 87p × 100

51 Sadiah buys three articles in a shop. They cost £1.99, 79p and £5.60. How much change should she be given from a £10 note?

52 Martin buys a clock-radio costing £17.20 from his catalogue. If he pays over a period of 20 weeks, how much is each weekly payment?

53 Find the cost of 1200 silicon chips at 63p each.

54 A pair of trousers, advertised at £16.90 in a sale, have been reduced by £1.90. What was the original price?

55 A firm pays its part-time workers £2.85 per hour. How much will a part-time worker earn in 12 hours?

56 Five removal men are given a tip of £6.75. How much should each one receive?

57 What would be the total cost of 12 boxes of crisps at £6.80 a box, plus 10 boxes of chocolate bars at £8.85 a box?

58 A coffee cup is reduced in price from 85p to 78p. How much would be saved on an order for 60 coffee cups?

59 Sarah is paid £81.90 for a 35-hour week. How much is she paid per hour?

60 Sam has a library book which is overdue by 24 days. The fine is 4p per day overdue. How much change will Sam receive from a £5 note when the fine is paid?

61 David has £98.30 in the bank. If he then writes a cheque for £101.80, by how much will he be overdrawn?

62 Omar buys 20 model cars for 50p each. He later sells them all for £12.50. How much profit has he made?

63 Julie bought a car for £4500 a year ago, but can get only £3850 for it now. How much has she lost?

64 Chris has £52.50 in the bank. He has five bills to pay: £13.60, £8.25, £17.19, £5.32 and £6.44. Is he still 'solvent', or is he 'in the red'? By how much?

65 Work out $(a)\frac{1}{3}$ of £12 $(b)\frac{1}{5}$ of £20 $(c)\frac{1}{10}$ of £15.

66 Work out $(a)\frac{2}{5}$ of £30 $(b)\frac{3}{4}$ of £14 $(c)\frac{5}{6}$ of £18.60.

67 Find $(a)\frac{7}{8}$ of £32 $(b)\frac{3}{10}$ of £38.60 $(c)\frac{9}{100}$ of £9.

68 'Second-hand car – £699 only.' If a deposit of $\frac{1}{3}$ is required, work out the deposit necessary to buy the car.

69 Edward's father has agreed to pay $\frac{3}{5}$ towards the cost of a set of golf clubs. How much will he have to pay towards a set which is priced at £95?

70 A particular dress material costs £8.37 per metre. How much would it cost Jill for $\frac{4}{9}$ of a metre of the material?

71 Work out (a) 5% of £3 (b) 60% of £4 (c) 20% of £5.20.

72 Work out (a) 75% of £8.80 (b) 10% of £3.70 (c) 12$\frac{1}{2}$% of £60.

73 Calculate the VAT at 15% payable on (*a*) a lamp costing £56, (*b*) a camera costing £80, (*c*) a microwave cooker costing £195.

74 Round off these amounts of money correct to the nearest penny (*a*) £4.627 (*b*) £5.018 (*c*) £1.999.

75 Round off these amounts of money correct to the nearest penny (*a*) £1.3241 (*b*) £92.035 (*c*) £6.6667.

76 Complete this bill:

    6 boxes of crisps   @ £6.80 =
12 boxes of chocolate @ £8.85 =
    8 boxes of sweets   @ £4.55 = _____

                           _____
              + VAT (15%) = _____
                    Total = _____

77 Complete this invoice:

3 boxes of bolts   @ £2.50 =
4 boxes of washers   @ £1.15 =
6 boxes of nuts   @ £1.43 = _____

                           _____
              + VAT (15%) = _____
                    Total = _____

78 Kevin receives 5% commission on sales of £2800. How much is this?

79 Workers in a factory are awarded a $2\frac{1}{2}\%$ bonus on profits. What bonus will be paid on profits of £10 250?

80 At a basketball game 35% of spectators were female. What percentage were male?

81 On one day, 28% of letters sorted at a post office were first class. What percentage were second class?

82 A surcharge of 12% was added to a holiday costing £360. What was the total cost of the holiday?

83 An order for 150 parts, each costing £3.80, qualifies for a discount of $12\frac{1}{2}\%$. Calculate the actual price paid for the order.

84 A builder offers a discount of 6% for any account settled in cash. Calculate the discount on an account of £2350.

85 Saleem buys a cooker for £280, but finds that it is scratched. If the shop will allow a discount of 20%, how much will the cooker cost Saleem?

86 Gerry is paid £350.55. What is this, to the nearest £?

87 Round off these amounts of money, correct to the nearest £.
(a) £3.04 (b) £18.90 (c) £139.52 (d) £123.45.

88 Round off these amounts of money, correct to the nearest £.
(a) £5.33 (b) £0.87 (c) £7.99 (d) £505.05.

89 A pop concert took ticket receipts of £12 632. Write this amount correct to the nearest £100.

90 Round off these amounts of money, correct to the nearest £100.
(a) £23 427 (b) £8090 (c) £40 060 (d) £4937.

91 Round off these amounts of money, correct to the nearest £10.
(a) £43 (b) £656 (c) £1908 (d) £7.85.

92 Barry buys a box of 20 alternators for a total of £150. He then sells them at £8.00 each. How much profit, or loss, does he make?

93 Jenny buys three typewriters, for £14.80, £16.20 and £10. After repairing them she sells them all for a total of £50. Find her profit.

94 A stereo can be bought for £280 cash, or by HP with a deposit of 25% plus 12 payments of £19.99. Find the difference between the cash price and the HP price.

95 A vacuum cleaner costs £112. It can be bought on HP by paying a deposit of $\frac{1}{8}$ plus 20 payments of £4.95. What is the difference between the cost price and the HP price?

96 Steve decides that he can afford to buy a new car, costing £6500, by putting down a deposit of £900, and paying 60 payments of £132.50 a month for a period of 5 years. Find the difference between the cost of the car and the total amount Steve will have paid.

97 Calculate the HP cost of a camera if the deposit on it is £20 and there are 12 payments of £7.60 each.

In questions 98 – 120 simplify the ratios as far as you can.

98	10:2	**99**	3:12	**100**	15:5
101	4:24	**102**	6:4	**103**	8:12
104	100:40	**105**	10:35	**106**	3:30
107	48:28	**108**	16:20	**109**	10:6
110	30:16	**111**	10:12	**112**	4:14
113	18:24	**114**	6:96	**115**	60:36
116	9:12	**117**	18:4	**118**	2:100
119	5:25	**120**	14:98		

121 If 10 felt-tip pens cost 99p, how much would 30 cost?

122 A pack of 5 photograph frames costs £4.49. How much would it cost for 30 photograph frames?

123 Six eggs cost 89p. How much would two dozen cost?

124 Three cans of limeade cost 59p. If I need 36 cans for a party, how much will they cost me?

125 Two loaves cost £1.06. How much would five loaves cost?

126 I decided to buy 4 plant pots, which cost me a total of 92p. How much would 3 have cost?

127 A group of 5 friends went to a cinema. The tickets cost a total of £16. Originally there were 7 people wanting to go to the cinema. How much would it have cost for 7 tickets?

128 If a packet of 120 notebooks costs £9.60, how much will it cost for 50 notebooks?

129 At a garden centre it cost me £7.50 for 10 tomato plants. My neighbour wanted 14 tomato plants. How much would she have had to pay?

130 How much will it cost for a piece of beef, weighing 2.74 kg, if 1 kg of beef costs £2.07? Give your answer to the nearest penny.

131 A length of electrical cable, 5.35 m long, cost me £3.47. About how much does one metre of cable cost?

132 'You can have that box of tangerines for £3.00.' If the tangerines weigh 6.58 kg, how much would a kilogram of tangerines cost? Give your answer correct to the nearest penny.

133 I made 30 bottles of wine from a kit which cost me £10.49. How much did it cost me to make one bottle of wine, correct to the nearest penny?

134 If each bottle in question 133 held 75 cl, how much would it have cost me to make one litre?

In questions 135–140, work out which is the better buy.

135 3 litres of creosote for £1.99 or 5 litres for £3.19.

136 12 eggs for 98p or 20 eggs for £1.71.

137 5 blank cassettes for £2.75 or 8 for £3.99.

138 600 g of sugar for 39p or 2.5 kg for £1.55.

139 30 metres of cable for £7.37 or 70 metres for £10.99.

140 12 oranges for £1.07 or 15 oranges for £1.30.

In questions 141–150, work out the value on the left-hand side of each formula.

141 $c = 24 + (5d)$ when $d = 3$

142 $T = 3H - 5$ when $H = 8$

143 $m = \dfrac{n + 6.4}{4}$ when $n = 13.6$

144 $p = \dfrac{50b - 17}{10}$ when $b = 3$

145 $K = 5(4t - 9.8)$ when $t = 6.7$

146 $A = P + (1.331n - 7.57)$ when $n = 4.5$ and $P = 12.50$

147 $s = \dfrac{2 \times 3.14 \times h}{3.16}$ when $h = 11.5$

148 $I = d + \dfrac{30}{d}$ when $d = 5$

149 $I = d + \dfrac{30}{d}$ when $d = 6$

150 $I = d + \dfrac{30}{d}$ when $d = 5.477$

151 The estimated cost given by a decorator for wallpapering a room is given by the formula: $C = K + (n \times £6.50)$, where C is the cost, K is a fixed charge of £10.00, and n is the number of rolls of wallpaper required.

     Calculate the cost of wallpapering a room which requires (*a*) 4 rolls, (*b*) 7 rolls, (*c*) 15 rolls.

152 To hire a pneumatic drill costs £5.00, plus £1.50 per day. Write a formula for the cost (C) of hiring a drill for x days. How much would it cost to hire the drill for (*a*) 3 days, (*b*) 10 days, (*c*) 2 weeks?

153 A formula for calculating travelling expenses is given by $E = £21.50 + (m \times £0.37)$, where m is the number of miles travelled.

     Calculate the travelling expenses due for a journey of (*a*) 12 miles, (*b*) 27 miles, (*c*) 100 miles.

154 If $T = D - (0.36 \times g)$, calculate the value of T when (*a*) $D = 23$ and $g = 100$ (*b*) $D = 72.8$ and $g = 84$.

155 Work out $234\,629 + 903\,628 - 186\,400 + 86\,046$.

156 I won £347 000 on the football pools, and paid £120 000 for a house, together with £15 000 for a car. How much have I left?

157 3.5 million watched an episode of a serial on television, and the next week 4.8 million watched the following episode. How many is this altogether?

158 In three weeks a store took £87 549, £76 945 and £106 384. What were the total takings for the three weeks?

159 'The number of unemployed people has dropped by one-tenth from 2 463 500.' How many are now unemployed?

160 4! means $4 \times 3 \times 2 \times 1 = 24$. Work out (*a*) 3! (*b*) 6! (*c*) 10! (*d*) 15! (*e*) 20!

5. Communication: numbers

Mathematics is an international language. It has been used to convey ideas of numbers from the earliest days of man. The earliest known artifact showing proof of counting is the Lembembo bone in Swaziland, which has been dated to about 35 000BC. Today numbers are used to pass on information, ideas, puzzles, and help us in our daily lives.

Egyptian number system

Number systems date back many years. The Egyptians used hieroglyphics to represent their numbers more than 5000 years ago.

| one | ten | 100 | 1000 | 10 000 |

These symbols were repeated when numbers were written, hence the number 23 457 is written as

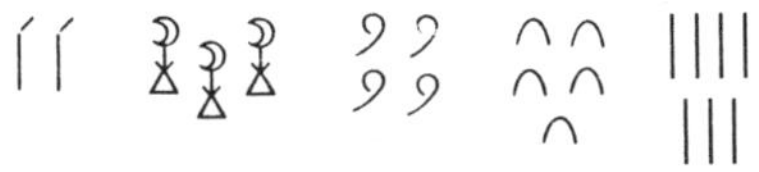

EXERCISE 5.1

Write down the numbers represented by these hieroglyphics.

1 
2 
3 
4

5 
6 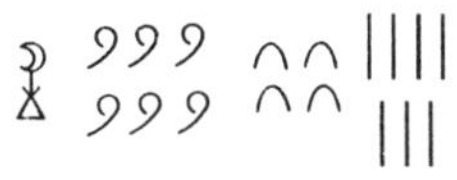
7

8 
9 
10

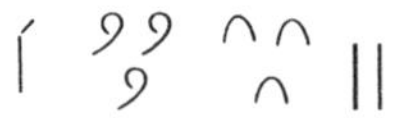

Write down the hieroglyphics which would represent these numbers.

11 52   **12** 44   **13** 451   **14** 362   **15** 7083

16 8234   **17** 17 000   **18** 23 521   **19** 30 304   **20** 10 437

Investigation A

The Chinese used the number system:

1	2	3	4	5	6	7	8	9

10	20	30	40	50	60	70	80	90

To represent the number 54 237 they would write:

Notice how 5 is picked from the first row of characters, 4 from the second row, 2 from the first row again, 3 from the second row and 7 from the first row, i.e. alternate rows.

7	8	3	9

Make up some numbers yourself, and ask a friend to decipher them.

Investigation B

Can you decipher this old Chinese pattern of numbers? Could you add another row to the diagram?

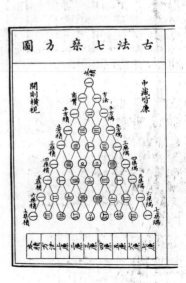

Roman number system

The Roman number system is based on symbols which can be combined to make up a series of numbers:

$$I=1 \quad V=5 \quad X=10 \quad L=50 \quad C=100 \quad D=500 \quad M=1000$$

The Roman number system is still used today and can be seen on clock and watch faces.

The numerals are usually written in order:

$$XVI = 10 + 5 + 1 = 16$$
$$XXXVII = 10 + 10 + 10 + 5 + 1 + 1 = 37$$

When the numerals are not written in order we have to do some subtraction:

$$IX = 1 \text{ before } 10 = 9$$
$$XIX = 10 + (1 \text{ before } 10) = 10 + 9 = 19$$

Roman numerals are frequently used to show the date on films and television programmes – see Example 1.

Example 1

© MCMLXXVIII (© means 'copyright')
Write down the number represented by this Roman numeral.

$$MCM = 1000 + (100 \text{ before } 1000) = 1000 + 900 = 1900$$
$$LXXVIII = 50 + 10 + 10 + 5 + 1 + 1 + 1 = 78$$

so the year is $1900 + 78 = 1978$.

EXERCISE 5.2

Write down the numbers represented by these Roman numerals.

1 XII	*2* XVI	*3* XXII	*4* XXXVI	*5* LVI
6 CCLXII	*7* LXVII	*8* CLXXVI	*9* IX	*10* XL
11 XC	*12* DCL	*13* CLXVI	*14* MMM	*15* MDCV

Write down the Roman numerals which would represent these numbers.

16 6	*17* 17	*18* 23	*19* 61	*20* 78
21 132	*22* 214	*23* 283	*24* 1330	*25* 1600
26 40	*27* 90	*28* 2452	*29* 1090	*30* 3243

Find the year during which the first of these programmes was broadcast.

31 Coronation Street © MCMLX

32 The Sooty Show © MCMLII

33 Andy Pandy © MCML

Write down in Roman numerals the date as it would be shown for these films.

34 Star Trek: The Motion Picture (1979)

35 Gone With the Wind (1939)

36 Ben Hur (1959)

37 E.T. the Extraterrestrial (1982)

38 How many Olympic games were there before the XXIIIrd games in 1984?

39 Olympic games occur every four years. What will be the Roman numerals for the next games?

40 When were the XIXth Olympic games?

Investigation C

Make a note of the programmes you watch one evening on television. Write down the title, type of programme and the date given at the end, noting as to whether it is given as Roman numerals or ordinary numbers. What type of programmes do you think use Roman numerals more than others?

Arabic number system

The Arabs learnt their mathematics from others, and their original number system was based on that used by Hindus from India.

Gradual changes took place in the way this was written, and by the sixteenth century it resembled the number system we use today:

Changing decimals to fractions

The Arabic number system was based on tens. Today we extend the number system to include decimals:

$$\ldots 1000\text{s} \quad 100\text{s} \quad 10\text{s} \quad \text{units} \quad \cdot \quad \tfrac{1}{10} \quad \tfrac{1}{100} \quad \tfrac{1}{1000} \ldots$$

A number such as 0.27 is represented by $0 \cdot \tfrac{2}{10} + \tfrac{7}{100}$

$$= \tfrac{20}{100} + \tfrac{7}{100} = \tfrac{27}{100} \quad \text{as you saw in Chapter 1.}$$

This is the way we compare decimals and fractions.

Example 2

$$13.25 = 13 \cdot \tfrac{2}{10} + \tfrac{5}{100} = 13\tfrac{20}{100} + \tfrac{5}{100}$$
$$= 13\tfrac{25}{100} = 13\tfrac{1}{4}$$

Remember to cancel your answer if you can.

EXERCISE 5.3

Copy and complete the table, writing each number under its correct column heading. The first one is done for you.

Decimal	1000s	100s	10s	units $\bullet$	$\frac{1}{10}$	$\frac{1}{100}$	$\frac{1}{1000}$
(a) 0.23				0 $\bullet$	2	3	
(b) 0.701							
(c) 0.205							
(d) 0.3							
(e) 59.04							
(f) 60.007							
(g) 23.42							
(h) 152.73							
(i) 13.31							
(j) 0.004							
(k) 6.8							
(l) 8.008							

EXERCISE 5.4

1 Some fractions are very common and we need to be familiar with their decimal equivalents. Copy and complete this table.

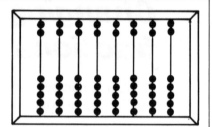

Decimal	Fraction
0.1	
0.01	
0.25	
0.5	
0.75	

2 Change into mixed fractions:
(*a*) 0.9   (*b*) 2.3   (*c*) 3.1   (*d*) 2.08   (*e*) 0.77
(*f*) 0.03   (*g*) 0.39   (*h*) 2.36   (*i*) 8.04

3 Change all the decimals in Exercise 5.3 into mixed numbers.

Changing fractions to decimals

$\frac{1}{4}$ means 'dividing a whole thing into 4 equal pieces and selecting one of these pieces'. So $\frac{1}{4}$ means $1 \div 4$.

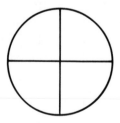

Try doing this on your calculator.

$$1 \div 4 = 0.25$$

By using a calculator to do this type of problem we can change fractions into decimals quite easily.

Example 3

$\frac{2}{5}$ means $2 \div 5 = 0.4$

EXERCISE 5.5

Use your calculator to change the following fractions into decimals:

1 $\frac{1}{8}$ 2 $\frac{3}{5}$ 3 $\frac{5}{8}$ 4 $\frac{1}{5}$ 5 $\frac{1}{20}$

6 $\frac{3}{8}$ 7 $\frac{4}{5}$ 8 $\frac{1}{16}$ 9 $\frac{5}{16}$ 10 $\frac{7}{32}$

11 $\frac{7}{8}$ 12 $\frac{21}{32}$ 13 $\frac{3}{16}$ 14 $\frac{17}{25}$ 15 $\frac{81}{100}$

16 $\frac{47}{1000}$ 17 $1\frac{1}{2}$ 18 $5\frac{7}{8}$ 19 $4\frac{3}{5}$ 20 $3\frac{5}{16}$

21 $5\frac{3}{100}$ 22 $2\frac{4}{25}$ 23 $8\frac{2}{5}$ 24 $4\frac{3}{4}$ 25 $10\frac{2}{5}$

Recurring decimals

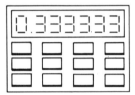

What is $\frac{1}{3}$ written as a decimal?

$\frac{1}{3}$ means $1 \div 3 = 0.333\,333\,3\ldots$

The 3s continue on for ever, but the calculator can only show some of them. This fraction cannot be shown exactly as a decimal, and the answer is known as a recurring decimal because it repeats itself, going on and on for ever.

EXERCISE 5.6

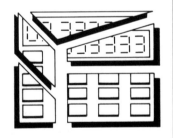

Change the following fractions into decimals:

1 $\frac{5}{6}$ 2 $\frac{2}{3}$ 3 $\frac{1}{6}$

4 (a) $\frac{1}{9}$ (b) $\frac{2}{9}$ (c) $\frac{3}{9}$

(d) By looking at your answers so far for ninths, can you predict what $\frac{4}{9}$ might be? Check it, and write down the decimal equivalents of $\frac{5}{9}, \frac{6}{9}, \frac{7}{9}$ and $\frac{8}{9}$.

5 (a) $\frac{1}{11}$ (b) $\frac{2}{11}$ (c) $\frac{3}{11}$

(d) By looking at your answers predict what $\frac{4}{11}$ might be. Then write down the decimal equivalents of $\frac{5}{11}, \frac{6}{11}, \frac{7}{11}$ and $\frac{8}{11}$.

Number patterns

In the exercise above there were patterns in the decimals you were producing as your answers. There are many more decimals with similar patterns of numbers.

$\frac{1}{7}$ means $1 \div 7 = 0.142\,857\,1$

Is this a recurring decimal?

Many calculators cannot show more than eight digits at one time, but if we use a calculator with a 10-digit display, or a computer, which can show more digits, we would get:

$1 \div 7 = 0.\underline{142}857\underline{142}\ldots$

We have a recurring decimal with a recurring sequence of numbers. What are the next numbers of the decimal answer? We would expect them to be 857 to continue the sequence, and the decimal would look like:

$0.142\,857\,142\,857\,142\,857\,142\,857\ldots$

We can clearly see the repeated pattern.

Example 4

Write $\frac{2}{7}$ as a decimal, and extend the pattern of numbers.

$\frac{2}{7}$ means $2 \div 7 = 0.\underline{285\,714\,2}$

so $\quad 2 \div 7 = 0.285\,714\,285\,714\,285\,714\ldots$

1 2 3 4 5
6 7 8 9 0
1 2 3 4 5
6 7 8 9 0
1 2 3 4 5
6 7 8 9 0
1 2 3 4 5
6 7 8 9 0
1 2 3 4 5
6 7 8 9 0
1 2 3 4 5
6 7 8 9 0
1 2 3 4 5
6 7 8 9 0

EXERCISE 5.7

Write the following fractions as decimals, extending your answers to show clearly the sequence of numbers.

1 $\frac{1}{7} = 0.142\,857\,142\,857\ldots$
$\frac{2}{7} = 0.285\,714\,285\,714\ldots$
$\frac{3}{7} =$
$\frac{4}{7} =$
$\frac{5}{7} =$
$\frac{6}{7} =$

Look at the numbers in all your answers. Can you say anything else about the patterns produced by all the decimal equivalents of sevenths?

2 $\frac{2}{13} = 0.153\,846\,153\,846\ldots$
$\frac{3}{13} =$
$\frac{4}{13} =$

Find the decimal equivalents of all the fractions of thirteenths, extending your answers to clearly show the sequence of numbers.

3 Copy these decimals and write in the next five digits for each of them.

(a) $\frac{1}{14} = 0.071\,428\,5714\ldots$ (b) $\frac{3}{14} = 0.214\,285\,714\ldots$

(c) $\frac{5}{14} = 0.357\,142\,857\ldots$ (d) $\frac{1}{21} = 0.047\,619\,047\ldots$

(e) $\frac{2}{21} = 0.095\,238\,095\ldots$ (f) $\frac{3}{21} = 0.142\,857\,14\ldots$

(g) $\frac{1}{26} = 0.038\,461\,538\ldots$ (h) $\frac{1}{13} = 0.076\,923\,076\ldots$

Investigation D

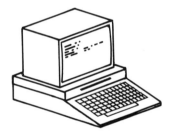

If you have a computer available, or a calculator which can display at least ten digits, investigate the fractions from $\frac{1}{2}, \frac{1}{3}, \ldots, \frac{1}{30}$ and their decimal equivalents. Find out which fractions do not produce recurring decimals. For those which do recur, make a note of the sequence of numbers which repeats. Does every recurring decimal have a pattern of numbers which repeats as a sequence of numbers?

Pattern numbers

In Book 1 we came across several groups of numbers which made up patterns of numbers.

Square numbers

1

$4 = 2 \times 2 = 1 + 3$

$9 = 3 \times 3 = 1 + 3 + 5$

What is the next square number? What do you notice about the numbers we add together to make up square numbers?

Triangle numbers

•   1   :· $3 = 1 + 2$ ·· $6 = 1 + 2 + 3$

What is the next triangle number?

EXERCISE 5.8

1 Find all the square numbers between 1 and 100.

2 Copy and complete:

$(1 \times 1) + (2 \times 1) + 1 =$
$(2 \times 2) + (2 \times 2) + 1 =$
$(3 \times 3) + (2 \times 3) + 1 =$
$(4 \times 4) + (2 \times 4) + 1 =$
$(5 \times 5) + (2 \times 5) + 1 =$
$(6 \times 6) + (2 \times 6) + 1 =$

What can you say about the numbers you have found?

3 This is the series of triangle numbers: 1, 3, 6, 10, 15 , ... Look closely at the numbers. How many are we adding on each time? Does this help you find the next triangle number? Find all the triangle numbers between 1 and 100.

Investigation E

This example of Islamic art can be reproduced in three different ways as shown. The star pattern is based on the square number 4. Can you produce the same type of patterns based on the square number 9, that is, starting with 9 stars?

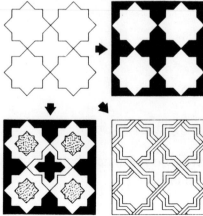

Number sequences

A sequence of numbers is a set of numbers that are connected by a relationship or a rule.

Example 5

Find the next three terms of the sequence $1, 3, 7, 13, \ldots$.

We need to discover a simple rule connecting the numbers.
    Find out how they have increased:

Can you see a simple relationship? For the next number we will add on 8, then 10, 12, etc. The next three numbers will be

$$13 + 8 \ = 21$$
$$21 + 10 = 31$$
$$31 + 12 = 43$$

Example 6

Find the next three numbers in the sequence $2916, 972, 324, \ldots$.

Investigate:

No clear rule there!

Investigate:

Here we have a simple relationship which works. The rule is to divide by 3, so the next three terms will be:

$$324 \div 3 = 108$$
$$108 \div 3 = 36$$
$$36 \div 3 = 12$$

To find a simple relationship connecting the numbers is largely a matter of trial and error until you find a rule which works throughout the sequence.

EXERCISE 5.9

State what you think should be the next three numbers in each of the following number sequences.

1 1, 3, 5, 7, ...
2 0, 2, 4, 6, ...
3 1, 4, 7, 10, ...
4 0, 3, 6, 9, ...
5 1, 5, 9, 13, ...
6 8, 15, 22, 29, ...
7 21, 30, 39, 48, ...
8 19, 32, 45, 58, ...
9 19, 38, 57, 76, ...
10 2, 4.5, 6, 7.5, ...
11 3, 6.5, 10, ...
12 1.3, 2.1, 2.9, 3.7, ...
13 2, 4, 8, 16, ...
14 3, 9, 27, ...
15 1, 5, 25, ...
16 3, 12, 48, ...
17 7, 21, 63, ...
18 0.001, 0.01, 0.1, ...
19 1, 2, 6, 24, ...
20 2, 10, 50, ...
21 2, 6, 18, 54, ...
22 1000, 500, 250, ...

Directed numbers

In Chapter 1 we were introduced to directed numbers, which were shown as minus numbers on the calculator. We also find directed numbers when we handle number sequences.

We are used to handling temperatures which are below freezing, i.e. minus temperatures. Freezing point is 0°C. What is the temperature one degree below freezing point? The thermometer tells us it is −1°C.

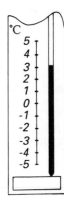

Example 7

The temperature is 3°C during the evening, but falls to −2°C during the night. By how many degrees has it fallen?

Counting down from 3°C to −2°C we get a fall of 5°C.

Example 8

The temperature changes from −2°C to 4°C. By how many degrees has it changed?

A *rise* of 6°C.

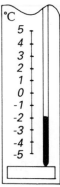

EXERCISE 5.10

Find the number of degrees the temperature has changed, stating whether it is a *rise* or a *fall* in temperature.

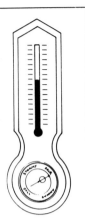

1 4°C to −2°C   **2** 6°C to −1°C   **3** −3°C to 3°C

4 3°C to −5°C   **5** −1°C to 4°C   **6** 5°C to 2°C

7 −4°C to −2°C   **8** 1°C to −1°C   **9** −3°C to −5°C

10 −1°C to −4°C   **11** 0°C to 5°C   **12** −5°C to −2°C

Find the new temperature in each case:

13 4°C falls by 2°C   **14** 1°C rises by 5°C   **15** −4°C rises by 6°C

16 −4°C falls by 2°C   **17** −4°C rises by 2°C   **18** 3°C falls by 6°C

19 −8°C rises by 8°C   **20** −1°C falls by 2°C   **21** −5°C falls by 2°C

22 −5°C rises by 8°C   **23** −1°C rises by 4°C   **24** 9°C falls by 8°C

The number line

Up the number line ◄ ► Down the number line

8 7 6 5 4 3 2 1 0 -1 -2 -3 -4 -5 -6 -7 -8

All numbers can be represented on a number line, which is really just like a thermometer on its side. Not only have we all the ordinary numbers, but the negative numbers as well. The number line can be used to help us with some number sequences.

Example 9

Complete the next three terms of these number sequences:

$5, 4, 3, 2, \ldots 1, 0, -1, \ldots$ goes down one each time
$8, 6, 4, 2, \ldots 0, -2, -4, \ldots$ goes down two each time
$-9, -6, -3, \ldots 0, 3, 6, \ldots$ goes up three each time

EXERCISE 5.11

Find the next three terms in each of the following number sequences:

1 $5, 4, 3, 2, 1, \ldots$ **2** $10, 7, 4, \ldots$

3 $12, 6, 0, \ldots$ **4** $8, 5, 2, \ldots$

5 $-7, -9, -11, \ldots$ **6** $-12, -9, -6, \ldots$

7 $8, 3, -2, \ldots$ **8** $17, 7, -3, \ldots$

9 $50, 40, 30, 20, \ldots$ **10** $-8, 7, -6, 5, \ldots$

11 $3, -9, 27, \ldots$ **12** $1, 2, 4, 7, 11, \ldots$

13 $7, 17, 30, 46, 65, \ldots$ **14** $3, 5, 4, 6, 5, 7, 6, \ldots$

15 $9, 13, 8, 12, 7, \ldots$ **16** $1, 1, 2, 3, 5, 8, 13, \ldots$

17 $1, 4, 9, 16, 25, \ldots$ **18** $3, 13, 22, 30, 37, ..$

19 $1, 2, 6, 24, \ldots$ **20** $5, 10, 25, 50, 85, \ldots$

Problem solving

These problems are a little different from the ones you have been asked to solve before. You can solve many of the problems using trial and error methods, but you will have to think carefully how you go about solving them, and a strategy might help you. Every problem is slightly different from the rest; no single method can be used for them all.

Example 10

Derek is laying a path using plain paving stones and coloured paving stones.

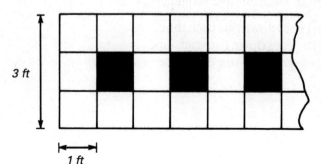

Length of path (ft)	1	2	3	4	5	6	7
Plain paving stones							
Coloured paving stones							

Complete the table for a path of length 7 feet.

You might need to look at the diagram to help you to start, but you should soon see a pattern in the numbers building up. Make the table longer, and find the number of each type of paving stone required for a path of 15 feet long. Can you continue the sequence of numbers? Could you find how many of each stone is needed for a path of length 30 ft?

EXERCISE 5.12

1 A path is being laid as in Example 10, but 5 stones wide as shown.

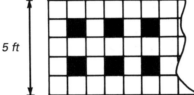

5 ft

Construct a table in which to place your figures, finding the number of each type of stone required for paths of lengths 1–8 feet. Watch for a pattern of numbers. How many of each stone would be needed for a path of length (*a*) 15 ft, (*b*) 20 ft, (*c*) 30 ft?

2 A single block has 5 faces showing. A column 2 blocks high has 9 faces showing. Find out how many faces are showing in a column 3 blocks high, 4 blocks high, etc. Do you notice a sequence of numbers building up? Can you predict the number of faces showing in a column of height (*a*) 10 blocks, (*b*) 15 blocks, (*c*) 20 blocks?

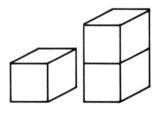

3 Toy boxes are to be stacked in a shop window as shown in the diagram. The number of boxes needed depends on how high the display is to be. By using the table below, find out the total number of boxes which would be needed for a display of height (*a*) 8 boxes, (*b*) 12 boxes, (*c*) 16 boxes.

Boxes high	1	2	3	
Total boxes	1	3		

4 The diagram shows the number of cans in each tier of a display in a shop. Notice how the cans are built up. Enter the information into a table.

1st tier   2nd tier   3rd tier   4th tier

Tier	1	2	3	
Cans	1	2		
Total cans	1	3		

Find the number of cans needed for a stack which is (*a*) 10 tiers high, (*b*) 15 tiers high.

5 The diagram shows a single block or step with 5 faces showing. Two steps can be made using three blocks placed together. This would have 12 faces showing. Three steps would need 6 blocks. Find out how many faces would be showing when we have 3 steps, 4 steps, etc., putting your results in a table.

Steps	1	2	3	
Blocks	1	3	6	
Faces	5	12		

  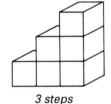

1 step   2 steps   3 steps

6 The diagram shows a house made out of dominoes. Complete and extend the table below to find the total number of dominoes needed for houses of different sizes.

Floors high	1	2	
Dominoes	3	3 + 5 = 8	

(*a*) How many dominoes would you need to build a house with (i) 6 floors, (ii) 10 floors, (iii) 15 floors?

(*b*) One box contains 28 dominoes. How many boxes would be needed to build each of the three houses described in (i), (ii) and (iii) above?

7 In diagram 1 there is one square. In diagram 2 there are 5 squares (4 small and the one whole square). How many in diagram 3? Draw a table to write in your findings. Extend the table to consider a 4th and 5th diagram. By looking at the pattern of numbers in your table, can you predict the numbers of squares to be found in a (*a*) 6th, (*b*) 7th, (*c*) 8th diagram?

 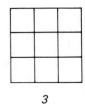

1   2   3

8 As in the previous question we have a series of diagrams made up of triangles. In diagram 1 there is one triangle, in diagram 2 there are 5 triangles (4 small and the one whole triangle). How many in diagram 3? Draw a table to write in your findings, and extend your table to consider more diagrams. Find the number of triangles to be found in a (*a*) 5th, (*b*) 6th diagram.

9 The diagram shows the number of runs and turns needed to get out of an expanding spiral.

> 1st stage: 2 runs, 1 turn; 2nd stage: 12 runs, 5 turns.

Draw a table in which you can enter your results, and extend the problem to the 3rd and 4th stages until you can see a sequence of numbers building up in the table. Using this sequence find the number of runs and turns needed for a 10th stage spiral.

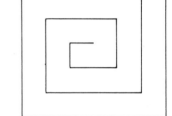

10 A man lives in a house across the river from his place of work. Whichever way he crosses the river he likes to come back home a different way.

With only one bridge he cannot come back a different way.

With 2 bridges he could cross bridge A and come back on bridge B, or cross bridge B and come back on bridge A: 2 different routes.

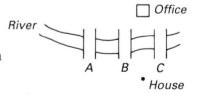

How many different routes could he use if he had 3 bridges to pick from? Enter your results into a table.

Bridges	1	2	3	4	5	
Different routes	0	2				

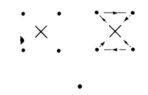

Complete the table and try to spot a sequence of numbers, then try to find how many different routes he would have with (*a*) 8, (*b*) 10, (*c*) 15 bridges.

11 You have to go to each point around you in turn. You must return to ✕ after you have been to any two points. What is the minimum number of moves you need to make to visit all the points and arrive back at ✕? 6 moves for 4 points.

Now try the same idea with 5 points. What is the minimum number of moves to arrive back at ✕?

Continue the problem for 6, 7, 8 points. Is there a sequence of numbers you could use? Find out the minimum number of moves for (*a*) 12, (*b*) 15, (*c*) 20 points.

12 There are three coins in a row: A, B, C. You want to exchange A
for C and C for A, that is the two end coins, but you can only
swap two adjacent coins at a time. What is the minimum number
of moves you must make?

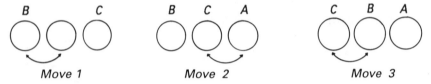

Try this again with 4 coins, then 5 coins, putting your results
in a table to help you identify the sequence of numbers. Predict
the number of moves needed when you have (*a*) 10, (*b*) 15,
(*c*) 20 coins.

Investigation F

A garage owner has 2 red and 2 blue cars he wishes to rearrange to form
a RBRB pattern which he thinks will be more appealing to customers. He
has one spare space which he can use to park a car while he moves the
other cars around. He can, of course, only move one car at a time. What is
the minimum number of moves he can make to rearrange the cars?

R	R	B	B	

What is the minimum number of moves he can make to rearrange the
cars if he has 3 red and 3 blue cars arranged RRRBBB and one free
parking space?

Set your results out in a table and extend the problem to consider more
cases.

Red cars	2	3	4	
Blue cars	2	3	4	
Minimum moves				

Can you see a quicker way to work it out for 5 cars of each colour?
You could extend this problem further by considering 3 red and 2 blue
cars, or 4 red and 3 blue, with the objective of getting a similar alternating
pattern.

Investigation G

Leaves of a book are numbered so that when assembled the pages are in order for reading:

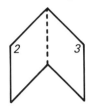 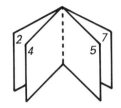 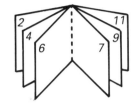

One page numbered 2,3; two pages numbered 2,7, 4,5; three pages numbered 2,11, 4,9, 6,7. Can you write down how the pages would be numbered for (*a*) 4 pages, (*b*) 6 pages, (*c*) 10 pages of a book?

Investigation H

A display for the Blackpool Illuminations is made up out of yellow (○) and blue (●) light bulbs. The display is assembled in stages as shown, each square to be lit in a repeated sequence fanning outwards. Draw a table in which to present your results.

Stage	Yellow bulbs	Blue bulbs	Total bulbs
1	0	1	1
2	8	1	9
3			

●

Stage 1

○ ○ ○
○ ● ○
○ ○ ○

Stage 2

● ● ● ● ●
● ○ ○ ○ ●
● ○ ● ○ ●
● ○ ○ ○ ●
● ● ● ● ●

Stage 3

Continue the table until you can find a sequence of numbers to help you answer the following questions.

How many (*a*) yellow and (*b*) blue bulbs go to make up (i) stage 5, (ii) stage 8, (iii) stage 12, (iv) stage 15?

Investigation I

A packaging machine is loaded with flavoured sweets in the order: strawberry, lime, lemon, orange, blackcurrant. They are put into packets containing 12 sweets. How many of each flavour of sweet will the first packet contain? The second packet? The third packet? How many packets will have been filled when they have used up an equal number of each flavour of sweet?

Extension

Consider how this sequence of packets is linked to the number of sweets; consider the same problem for a packet containing 13, or 14 sweets, and see how the answers to the above questions are different.

Investigation J

A knight on a chess board moves two squares in one direction, turns 90° left or right, and then moves on one more square:

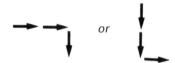

Starting at one corner of a board of 5 × 5 squares, find out how many moves it takes to get to the opposite corner. Then try the same problem for a 6 × 6 board, etc.

Extension

Consider a simple game: two opponents, you and your friend each try to get to the opposite corner of the board first without meeting. Can you decide on a strategy to help you win?

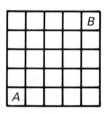

6. Communication: graphics (1)

Graphs and charts

Much information in magazines, newspapers, on television, advertising boards and in transport offices is given by means of graphs and charts. Most are clear and give you a fair picture; however, some can be misleading – intentionally so in some cases. Therefore it is as well to take particular care when interpreting information presented graphically.

To begin with here are some easy graphs to read. Take care, however, when there is a scale to read – each square does not always stand for one unit. For example, the diagram in question 1 below shows two different scales, which you need to read carefully. Most mistakes in graphical problems are made by not reading the scales correctly.

EXERCISE 6.1

1 A large flask of water is heated to 50°C. It is then allowed to cool, and the temperature is taken every two minutes. The diagram shows the temperature of the water up to ten minutes after starting to cool.

(*a*) What is the temperature after 6 minutes?

(*b*) What is the temperature after 2.5 minutes?

(*c*) How many minutes does it take to cool to 46°C?

(*d*) How many minutes does it take to cool to 42°C?

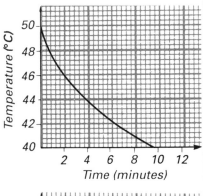

2 Mrs Wilson has run a mail-order catalogue since 1983. The diagram shows her commission each year.

(*a*) Estimate her commission in 1983.

(*b*) What was her commission in 1985?

(*c*) What was her total commission in the years 1984, 1985 and 1986?

(*d*) Make a guess at her 1989 commission.

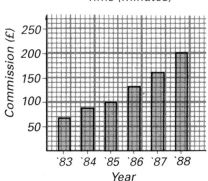

3 The diagram shows the percentage of income earned from the different vegetables that a farmer grows.

(*a*) Estimate the percentage earned from broad beans.

(*b*) Estimate the percentage earned from runner beans and leeks.

(*c*) What percentage is earned from sprouts and peas?

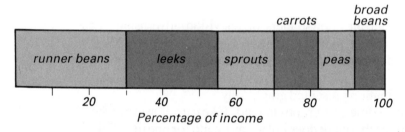

4 As part of her project in humanities, a girl measured the amount of rainfall in each month from January to December. The diagram shows the rainfall, measured in millimetres.

(*a*) How much rain fell in September?

(*b*) Which was the wettest month?

(*c*) Which was wetter, the first six months or the last six months?

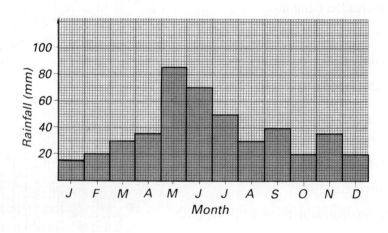

5 The number of horses, dairy cattle, beef cattle and sheep on a
large collection of farms in Russia is shown in the diagram.

(*a*) How many horses are there?

(*b*) How many cattle are there? (Beef and dairy.)

(*c*) 'There are more sheep than any other animal.'
Is this statement correct?

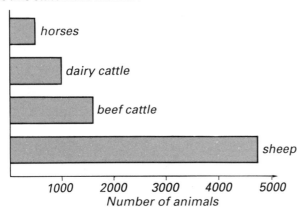

6 The diagram shows the actual, and estimated, population of
Ireland for the last 120 years or so.

(*a*) What was the actual population in 1901?

(*b*) What was the estimated population for 1961?

(*c*) By how many did the actual population drop from 1881 to
1981?

(*d*) In 1981, by how many was the estimated population greater
than the actual population?

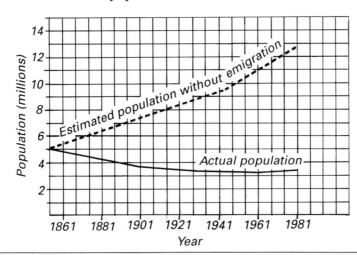

7 A patient's temperature chart is shown for a period of 24 hours.

   (*a*) What was his temperature at 0800?

   (*b*) What was his temperature at 1800?

   (*c*) Estimate his temperature at 1300.

   (*d*) What was his highest temperature?

   (*e*) By how much did his temperature drop from 1400 to 2000?

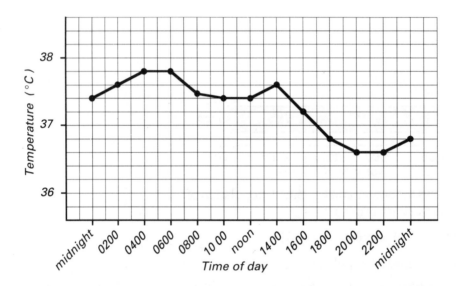

8 A council distributes its industrial grant to five areas, as shown in the diagram.

   (*a*) Which area is given the largest proportion of the grant?

   (*b*) What percentage is given to services and distributive trades together?

   (*c*) How much, out of a grant of £100, should go to transport and communication?

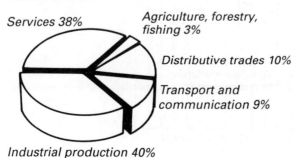

9 Colin and Rachel, as part of an enterprise scheme, sell writing paper. Their takings for a week are shown in the diagram.

(*a*) How much did they take on Thursday?

(*b*) Which was their worst day?

(*c*) How much did they take in the week?

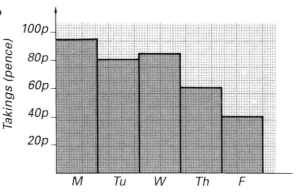

10 The conversion graph of temperatures between Fahrenheit and Celsius is shown in the diagram.

(*a*) What is 20°C in °F?

(*b*) What is 100°F in °C?

(*c*) Freezing point is 0°C. What is this in °F?

(*d*) The temperature goes up from 15°C to 25°C. What rise is this in °F?

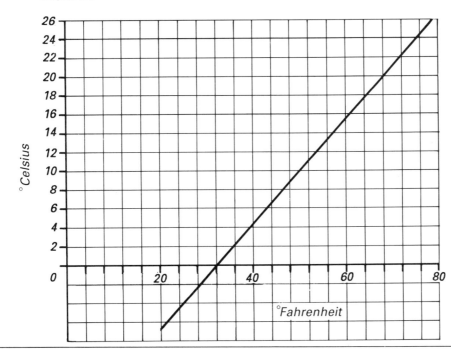

Bar graphs

Information is often displayed by means of a bar graph, as in questions 2 and 5 in Exercise 6.1. The bars can be drawn either vertically or horizontally, either joined up or with a gap between each bar.

Example 1

The bar graph shows the number of hours spent per week on subjects studied by the third year at a comprehensive school.

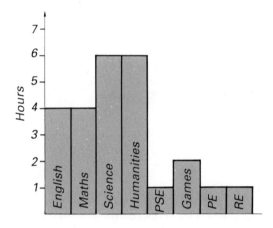

(*a*) How many hours are spent on English?

(*b*) How many hours are spent on science and humanities together?

(*c*) Is more time spent on maths than on PE and games?

(*a*) There are 4 hours spent on English.

(*b*) There are $6 + 6 = 12$ hours spent on science and humanities.

(*c*) 4 hours are spent on maths, and altogether 3 hours are spent on PE and games; thus *more* time (1 hour) is spent on maths than on PE and games.

EXERCISE 6.2

1 A die is thrown 50 times; the diagram shows how often each number turned up.

(*a*) How many times did a 4 turn up?

(*b*) How often did an odd number occur?

(*c*) 'A * turned up twice as often as *.' Replace each asterisk with the correct number.

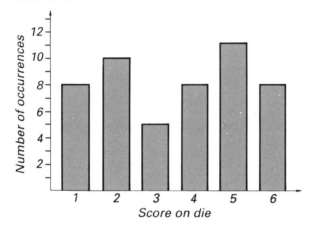

2 The diagram shows the area of each of the 'Great Lakes'.

(*a*) Which two lakes have about the same area?

(*b*) What is the area of Lake Erie?

(*c*) What is the total area of all five lakes?

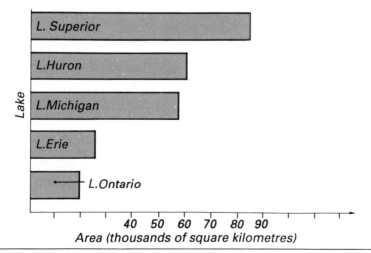

3 The amount of the mineral bauxite which was produced in 1981 is shown in the diagram.

(*a*) How many tonnes were produced in South Africa?

(*b*) Estimate the total bauxite production in 1981.

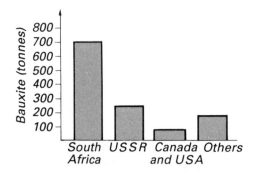

4 The diagram shows the average price of oil, in dollars per barrel, during each year from 1972 to 1980.

(*a*) What was the price in 1975?

(*b*) The price stayed the same for two successive years. Which two?

(*c*) What was the increase in price from 1973 to 1974?

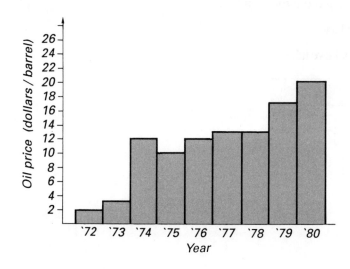

5 The highest temperature in various European places during one February is shown in the diagram.

(*a*) Which was the warmest place?

(*b*) Which places had a highest temperature of 13°C or more?

(*c*) Three places had the same temperature. Which were they?

(*d*) How much warmer was it in Lisbon than it was in Manchester?

(*e*) What was the difference in temperature between the warmest and coolest places?

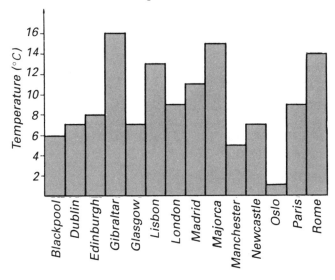

6 The diagram shows how many therms of gas Ms Hargreaves used to heat her flat during one week.

(*a*) How many therms did she use at the weekend (Saturday and Sunday)?

(*b*) How many therms did she use altogether in the week?

(*c*) She shut off the heating in one room on one day. Which day?

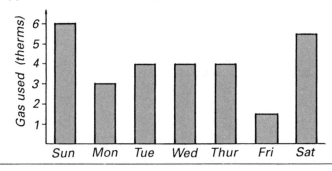

7 The unemployment rate during 1984 in five northern areas is shown in the diagram.

(*a*) Which area had most unemployment?

(*b*) What percentage was this?

(*c*) Which two areas had about the same unemployment rate?

(*d*) What was Northumberland's unemployment rate?

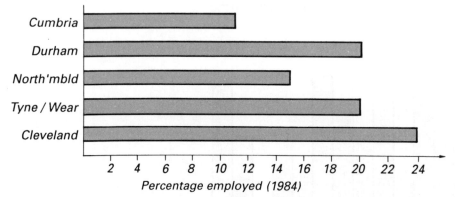

Percentage employed (1984)

8 The rate of inflation in various countries in 1975 is shown in the diagram.

(*a*) Which country had 10% inflation?

(*b*) What was the UK's inflation rate?

(*c*) Which country had just under half of the inflation rate of Belgium?

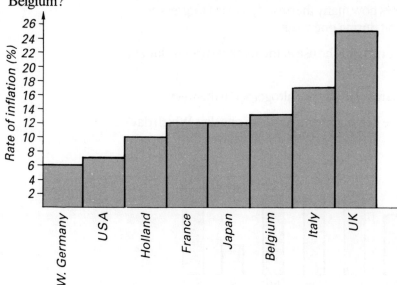

9 The amount that each share in a particular company earned during recent years is shown in the diagram.

(*a*) How much did each share earn in 1984?

(*b*) Which year yielded the lowest return?

(*c*) How much would your 1000 shares have earned for you in 1986?

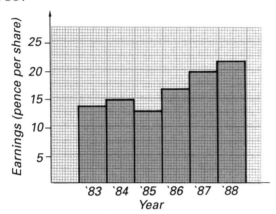

10 As part of an assignment Jan makes a note of the colours of cars in a supermarket car park.

(*a*) How many cars were blue?

(*b*) How many cars were either red or green?

(*c*) The car park can take a maximum of fifty cars. How many more cars could the car park take?

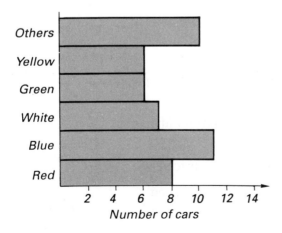

Pie charts

Quite often a pie chart is used to convey information and it gives a quick visual comparison, usually of four or five things. Sometimes the pie chart circle is designed as a £1 coin to show how much money, out of every pound, is allocated to the various sectors. These values will be the same as the **percentage** allocations, because percentage is the same as 'pence in the £'.

Example 2

The diagram compares the four main fuels used in Great Britain in 1977.

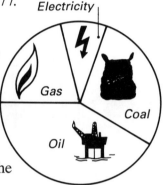

(*a*) Which fuel was used most?

(*b*) Two fuels together made up about a third of the total. Which two?

(*c*) 'About twice as much oil was used as gas.' Is this correct?

(*a*) The largest sector represents *oil*.

(*b*) *Electricity* and *coal* made up about a third.

(*c*) Yes, because the sector representing oil is about twice the size of the sector representing gas.

EXERCISE 6.3

1 The pie chart shows how much of a company's goods are sold in different parts of the world.

(*a*) Where are most sales made?

(*b*) Which two places have about the same sales?

(*c*) If sales to Asia amount to £1 000 000, estimate the sales to Europe.

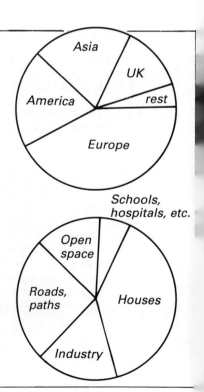

2 Land used in a typical South American city is shown in the pie chart.

(*a*) What fraction, approximately, of land is taken up with roads, paths, etc?

(*b*) 'About 60% of land is used for houses.' Is this correct?

(*c*) 'Houses take up as much land as roads, paths and open spaces.' Is this correct?

3 The pie chart shows the destinations of third-year pupils who went on holiday to Europe.

(*a*) Which country was most popular?

(*b*) Which was next in popularity?

(*c*) If 50 pupils went to Spain, estimate how many went to Italy.

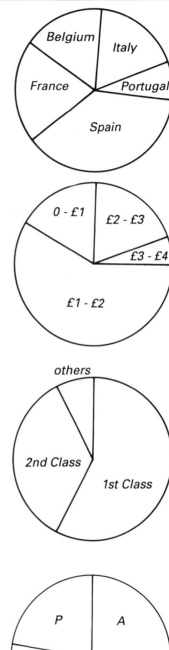

4 Sixty pupils were asked how much pocket money they were given. The pie chart shows the information.
Which of these statements are true, and which are false?

(*a*) 'About 10 pupils had £2 – £3.'

(*b*) '27 pupils had £1 – £2.'

(*c*) 'About 45 pupils had less than £2.'

5 On one Friday, my local post office sold 100 postage stamps. From the pie chart, estimate the number of first class, second class, and other stamps that were sold on that Friday.

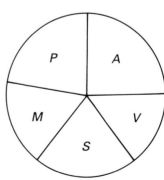

6 The pie chart shows how the cost of a record is made up.
P = profits/costs of record company
A = fees for artists/composers
M = manufacturing/distribution costs
S = profit for shop
V = VAT
Assuming that the record costs £6.99, estimate

(*a*) how much is profit (P + S),

(*b*) the artist's/composer's fees,

(*c*) the VAT.

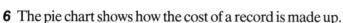

7 The pie chart shows the ratio of the number of players in each section of an orchestra.
   If there are 25 string players, estimate

(*a*) the number of woodwind players,

(*b*) the number in the percussion section,

(*c*) the total number of players in the orchestra.

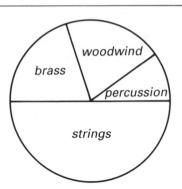

8 Anne, Brenda, Caroline and Denise have ten pets between them. The pie chart represents the number of pets belonging to each girl. (A = Anne's pets, etc.)

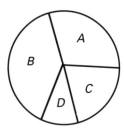

(*a*) If Denise has just one pet, how many do each of the other girls have?

(*b*) The sectors of the pie chart are to be rearranged to form two semicircles. Copy and complete the diagram, writing the letters A, B, C and D in the appropriate sector.

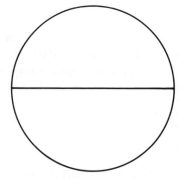

9 The pie chart shows the amount of four types of sugar sold by a supermarket in one week. Altogether the supermarket sold 500 kg of sugar in the week.

(*a*) Which types had about equal sales?

(*b*) About how much granulated sugar was sold?

(*c*) About what percentage of sugar sales were of icing sugar?

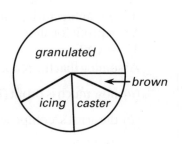

10 In a text processing examination, you can receive either a pass, a credit or a distinction; otherwise you fail.

The pie chart shows the result of an examination entered by 85 candidates. Estimate how many candidates there were at each grade.

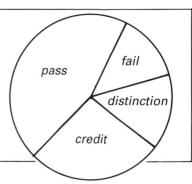

Timetables – bus and train

You will find increasingly that you will need to be able to understand timetables of buses and trains. The buses and trains may still be late, but you will feel better if you can find out from the timetable what time they are supposed to arrive!

EXERCISE 6.4

1 I catch the school bus at Station Road.

(*a*) How long does it take to reach Redshaw Avenue?

(*b*) Jeff gets on the bus at Redshaw Avenue. How long is his bus journey to school?

(*c*) Assuming that the bus is 9 minutes late, will I miss the 9.00 a.m. bell?

Station Rd	8.28
Redshaw Ave.	8.41
Post Office	8.48
School	8.53

2 The times of the first six trains in the morning from Newcastle to London are shown in the table.

(*a*) What time does the 0652 arrive in London?

(*b*) Olive meets her mother's train which arrives at 1150. What time did it leave Newcastle?

(*c*) How long, in hours and minutes, does the 0735 take for the journey?

INTERCITY

Newcastle	London
Newcastle depart	London arrive
0525	0950
0550	0859
0645	1033
0652	0955
0735	1054
0834	1150

Timetable © British Rail. This timetable is not current.

3 A special Sunday bus goes from Halifax to Nottingham.

(*a*) How long does it take to go from Halifax to Dewsbury?

(*b*) Farouk joins the bus at Wakefield. He is going to Chesterfield. For how long is he on the bus?

(*c*) How long is the whole journey from Halifax to Nottingham?

```
Halifax......0930
Dewsbury.....1010
Wakefield....1030
Barnsley.....1050
Sheffield....1120
Chesterfield.1150
Nottingham...1300
```

4

Dover (depart)	0150	0405	0600	0750	1005	1200
Calais (arrive)	0335	0550	0745	0935		1345

The times of the morning ferry from Dover to Calais are shown above.

(*a*) How long does it take the ferry to make the crossing?

(*b*) What time does the 1005 arrive at Calais?

(*c*) If I arrive in Dover at 0340, how long will I have to wait before the next ferry leaves?

5 A new hovercraft service is proposed to go twice daily from Liverpool to Belfast, calling at Douglas in the Isle of Man on the way.

Liverpool	(depart)	0900	1530
Douglas	(arrive)	1040	1710
Douglas	(depart)	1055	1725
Belfast	(arrive)	1230	1900

(*a*) How long is the journey from Liverpool to Douglas?

(*b*) How long does the hovercraft spend in Douglas?

(*c*) How long is the journey from Douglas to Belfast?

(*d*) If an extra service was suggested which was to arrive at Belfast at midnight, at what time should it leave Liverpool?

6

	X	Y	Z
Snake Station	1400	1515	1630
Lion Lane	1412	1527	1642
Dingo Drive	1419	1534	
Tiger Terrace	1432		1702
Rhino Road		1608	1723
Snake Station	1500	1615	1730

The miniature train in Funland Holiday Park makes three trips each afternoon. You can buy a ticket for trip *X*, trip *Y* or trip *Z*.

(*a*) How long is the journey from Snake Station to Dingo Drive?

(*b*) If I join the train at Lion Lane and leave it at Rhino Road, for how long will I be on the train?

(*c*) How long is each trip?

(*d*) If Steve arrives at Lion Lane at 3.33 p.m., by how many minutes has he missed trip *Y*?

(*e*) There is a missing time on each trip. Work out what each missing time should be.

7

Central Station	0645
Grange Road	0651
Piper Drive	0654
Green Lane	0702
Waterloo Hall	0711
Bruntley	0728

The first bus in the morning from Central Station to Bruntley leaves at 6.45 a.m. as shown in the table. After that a bus leaves at 45 minutes past each hour until the last one at 9.45 p.m.

(*a*) What time will the 0845 from Central Station arrive at Green Lane?

(*b*) How long does it take to travel from Grange Road to Waterloo Hall?

(*c*) Mrs Sanderson, who lives in Piper Drive, is meeting her friend Mrs Grant in Bruntley at 1330. At what time should she catch the bus?

(*d*) What time does the last bus arrive at Bruntley?

8 A copy of the weekday timetable for the Settle–Carlisle railway line during 1987 is given below.

SETTLE	0626 0942 1146 1421 1729
Horton	0637 — 1156 1432 1740
Dent	0654 1006 1213 1449 1757
Garsdale	0700 1012 1219 1455 1803
Kirkby Stephen	0720 1031 1239 1515 1823
APPLEBY	0733 1044 1253 1529 1836
Langwathby	0748 — 1308 1545 1851
Lazonby	0754 — 1315 1552 1858
Armathwaite	0802 — 1323 1600 1906
CARLISLE	0821 1127 1344 1622 1927

(The 0942 from Settle does not stop at those stations marked '—'.)

(*a*) If I catch the 1421 from Settle, when should I arrive in Dent?

(*b*) How long does the 1729 train take to travel from Settle to Carlisle?

(*c*) Claire arranges to meet Susan at Carlisle Station at 2.00 p.m. What time train should Claire catch from Garsdale?

(*d*) Which train (1st, 2nd, 3rd, 4th or 5th) takes a minute less to travel from Garsdale to Kirkby Stephen than the other four?

(*e*) How long does the first train in the morning take to travel from Langwathby to Carlisle?

9 The North Wales Coast Scenic Line train times are shown on page 97. The first table shows the morning times out from Chester, and the second table shows the return times back in the evening.

(*a*) How long does the 1207 from Chester take to travel to Holyhead?

(*b*) If I catch the earliest train that goes to Holyhead, and I return by the latest train in the evening, how long can I spend in Holyhead, to the nearest hour?

(*c*) If Ivor can get to Bangor Station by 5.30 p.m., will he be able to be in Chester by 7.10 p.m.?

(d) Len goes to Holyhead on the 1105 from Llandudno
Junction, and returns there on the 1910 from Holyhead.
Which train journey is longer?

Chester	0952 1007 1022 1052 1122 1207
Rhyl	1028 1042 1054 1128 1156 1328
Llandudno Junction	1050 1105 1116 1151 1217 —
Bangor	— 1127 — 1210 — 1326
Holyhead	— 1158 — 1248 — 1358

Holyhead	1615 — — 1705 — — 1910
Bangor	1643 — — 1733 — 1850 1947
Llandudno Junction	1707 1713 1743 1757 1813 1913 2013
Rhyl	1729 1734 1809 1819 1834 1934 2034
Chester	1802 1811 1847 1900 1911 2011 2111

(The trains do not stop at stations marked '—'.)

10 The times of three trains from Stevenage to King's Cross at
about lunchtime are shown in the table.

Stevenage	1151 1221 1251
Knebworth	1155 1225 1255
Welwyn North	1159 1229 1259
Welwyn Garden	1203 1233 1303
Hatfield	1207 1237 1307
Potters Bar	1213 1243 1313
Finsbury Park	1224 1254 1324
King's Cross	1229 1259 1329

(a) How long does the 1151 take to travel from Stevenage to
Hatfield?

(b) Mandy arrives at Knebworth Station at 1214. What is the
earliest time that she can arrive at Finsbury Park?

(c) Derek arrives at Hatfield Station at 1248. How long does he
have to wait for the next train to Potters Bar?

(d) Between which two stations is the journey time stated to be
exactly half an hour?

(e) The next train starts from Stevenage at 1326. Make a table
of the times at which it should arrive at all the stations
between Stevenage and King's Cross.

Coordinates

You have no doubt come across situations where the position of some point, as measured from another point, is required. A knight's move in chess, determining a 'hit' in the game of 'battleships', and stating a grid reference on a map are all examples of identifying a position from some starting point.

In case you have not met coordinates before, let us see what is needed in order to locate a point.

(*a*) We need a starting point, or 'origin' (call it 0)

(*b*) We need a line *across* the page, with distances from 0 marked on it (centimetres will be convenient).

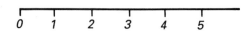

(*c*) We need a similar (perpendicular) line *up* the page, again with distances from 0 marked on it.

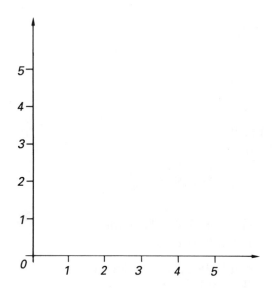

With these three things (an origin, and two lines) we can identify the position of any point on the paper, by stating its distance *across* the page, then its distance *up* the page. These two numbers are called the **coordinates** of the point.

Example 3

What are the coordinates of the points A, B and C in the diagram?

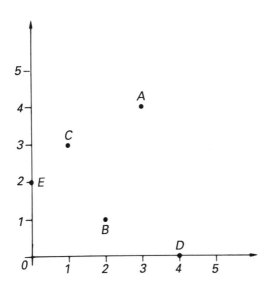

A is 3 cm *across* and 4 cm *up* from the origin, so its coordinates are 3 and 4. We write coordinates in brackets, with a comma, like this: (3,4).
   Similarly, B has coordinates (2,1) and C has coordinates (1,3).
   For convenience, we often say that 'A is the point (3,4)', or even 'A is (3,4)'. So we can say that B is (2,1) and C is (1,3).

Example 4

State the coordinates of the points D and E in the diagram in Example 3.

As D is 4 cm *across* and 0 cm *up*, D is the point (4,0). Similarly, E is 0 cm *across* and 2 cm *up*, so E is (0,2).

In Exercise 6.5 the diagrams may be scale drawings, in order to get them on the page. This will make no difference to the way in which the coordinates are worked out, or written down.

EXERCISE 6.5

1 What are the coordinates of the points P,Q,R,S and T in the diagram?

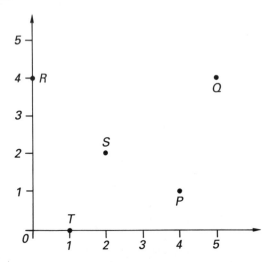

2 Write down the coordinates of M and N. If J is the middle point of the line joining M and N, what are its coordinates?

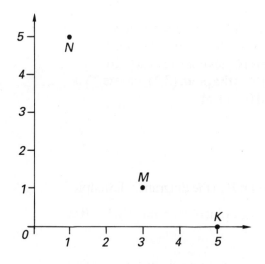

3 K is (5,0) and M is the middle point of the line joining K to L. What are the coordinates of the point L?

4 It is possible to draw a straight line through the three marked points on the diagram. Write down the coordinates of any other points on this line.

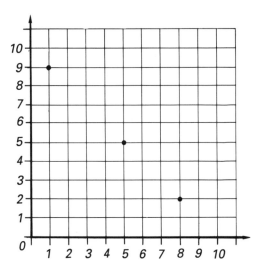

5 Write down the coordinates of the points G, H, J and K, which represent wooden pegs fixed into the ground. If a string is tied from G to H, and another is tied from J to K, what are the coordinates of the point where the strings cross each other?

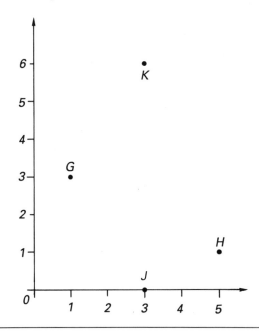

6 W, X and Y are three corners of a square.

  (*a*) Write down the coordinates of W, X and Y.

  (*b*) What are the coordinates of Z, the fourth corner of the square? How could you have worked out the coordinates of Z from knowing only the coordinates of W, X and Y?

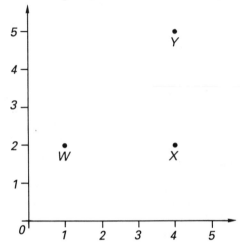

7 To reach the point T from point B you have to go 4 cm *across* and 3 cm *up*.

  (*a*) What are the coordinates of B?

  (*b*) What will be the coordinates of T?

  (*c*) If instead, B had been the point (2,0), what would the coordinates of T have been?

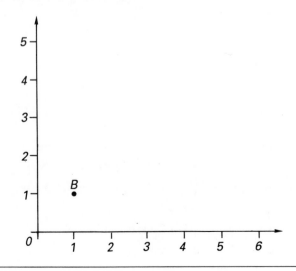

8 The points in the diagram could be joined up to form the letter 'L'.

(*a*) Which extra point would be needed to form the letter 'K'?

(*b*) A further point would be needed for the letter 'H'. Which point?

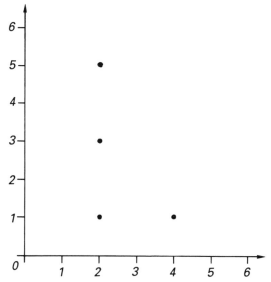

9 Figures on a digital watch are usually formed by joining some or all of the six points in the diagram.

(*a*) Which point would *not* be used in forming the figure '7'?

(*b*) Which point would *not* be used in forming the figure '6'?

(*c*) Which of the figures from 0 to 9 require all six points?

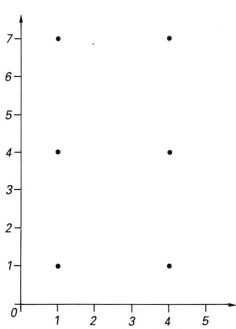

10 Write down the coordinates of some of the points lying on the sloping line in the diagram.

What can you say about the coordinates of *all* points which lie on the sloping line in the diagram?

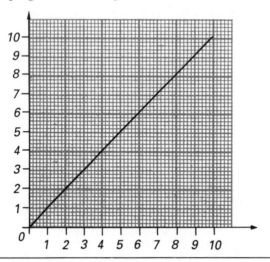

It can be helpful to give a name to the two lines which we draw through the origin.

The line *across* the page is the '*x* axis'.
The line *up* the page is the '*y* axis'.

The coordinates therefore can be identified separately.

The *first* number is the *x* coordinate, and the *second* number is the *y* coordinate.

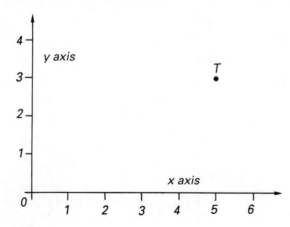

If we take the point T(5,3), for example, then the *x* coordinate is 5 and the *y* coordinate is 3.

As well as being able to read off the coordinates of points in a diagram, you need to be able to mark points whose coordinates you are given.

We sometimes use the phrase 'plot the points' – this just means mark the points on the diagram.

Example 5

Mark the points R(4,7), S(6,4), T(8,1), U(2,3) and V(4,0) on a grid with *x* and *y* axes.

(*a*) Which three points lie on a straight line? Draw the line.

(*b*) Draw a line joining U to T and another joining S to V.
What are the coordinates of the point where these lines meet?

(*a*) The points R, S and T lie on a straight line.

(*b*) UT and SV meet at (5,2)

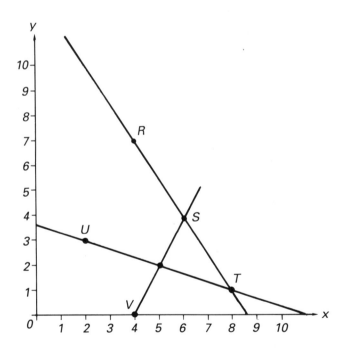

In Exercise 6.6, you may need to draw each axis up to 10 units long. Most questions, however, will require you to plot points with each coordinate less than 10.

EXERCISE 6.6

1 Plot the points A(3,2) and B(6,4) on a diagram. Join them up with a ruler, and continue the line back. Does it pass through the origin (0,0)?

2 The points (8,2), (8,6), (2,6) and (2,2) form a rectangle.

(*a*) Plot the points and draw the rectangle.

(*b*) Join the diagonals. What are the coordinates of the centre of the rectangle?

3 Plot the points (0,4), (2,4), (3,4), (5,4) and (8,4). Join them up.

(*a*) What do you notice about the *y* coordinate of each point? We call this line '*y* = 4'.

(*b*) Write down the coordinates of some more points which lie on this line.

(*c*) Draw, on your diagram, the line where all points have a *y* coordinate of 9. Write '*y* = 9' along the line.

4 Plot the points (7,0), (7,1), (7,4), (7,7) and (7,10). Join them up.

(*a*) What do you notice about the *x* coordinate of each point? We call this line '*x* = 7'.

(*b*) Write down the coordinates of some more points which lie on this line.

(*c*) Draw, on your diagram, the line where all points have an *x* coordinate of 2. Write '*x* = 2' along this line.

5 Plot these points, and join them up, *in order*, with straight lines: (3,1), (8,1), (8,3), (5,3), (5,7), (3,7) and back to (3,1). What letter is shown?

6 Plot the points (1,1), (1,4) and (9,4).

(*a*) These are three corners of a rectangle. What are the coordinates of the fourth corner? Mark it on your diagram.

(*b*) If instead the fourth corner was at (9,7), what shape would you have? Draw it on your diagram.

7 Do the points (3,0), (9,2), (7,8) and (1,6) form a square? Plot them, and check by measuring the angles and sides.

8 A knight's move in chess is 2 squares forward in one direction followed by 1 square either to the left or to the right. For example, if the knight was at (6,5) then it could move to (8,6). (Assume the coordinates are at the centre of a square on the chess board.)

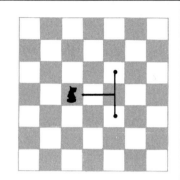

(*a*) Mark with a 'K' the point (6,5).

(*b*) Mark *all* of the possible points that the knight could move to, including (8,6).

(*c*) Write down the coordinates of these points.

9 Plot the points A(0,8), B(4,0) and C(8,8), and join them up to form a triangle. There is a point X inside the triangle where the lengths XA, XB and XC are all the same.

(*a*) What are the coordinates of X?

(*b*) What is the length of XA (or XB or XC)?

10 Join the points P(0,9), Q(7,9), R(7,7), S(9,7), T(9,1) and U(0,1). This represents a large classroom (1 unit = 1 metre).

(*a*) The teacher's desk has corners at (1,2), (3,2), (3,3) and (1,3). Draw its position on the diagram.

(*b*) There are fifteen light bulbs in the room. Each one is 2 metres from the next one either across or down the room. One is at (1,8). Mark the position of the others with small crosses on your diagram (two of the lights are directly at the corners of the teacher's desk).

(*c*) Square tables, of side 1 metre, are to be arranged in the room. They must be at least 1 metre away from each other, but may be placed touching sides PU or PQ. They must, however, be at least 1 metre from QR, RS, ST and TU. What is the largest number of tables that can be fitted in the room? Show your answer on your diagram.

Investigation

Using scales up to 10 on each axis, write the first letter of your name. You can make the letter simply by drawing straight lines, or by making it 1 cm or 2 cm thick (see Exercise 6.6, question 5). Now write down the coordinates of the points which will make your letter.

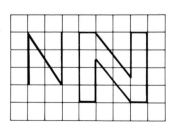

If you miss out the brackets, you will have a string of numbers which could be a code for your letter. You can extend this to your initials if you wish, or even your full name.

Extending the axes

It is often necessary to identify points which are below the x axis, or to the left of the y axis.

In order to do this we simply extend each axis, and label each with negative numbers, as in the diagram in Example 6. In order to be even clearer, we sometimes label the positive numbers with a '+' sign.

Example 6

What are the coordinates of the points A,B,C,D,E and F in the diagram?

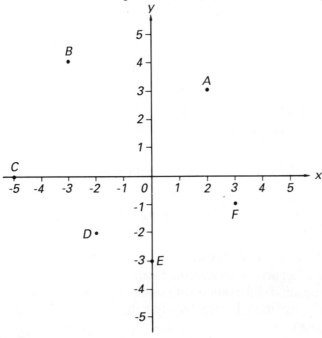

A is (2,3), B is (−3,4), C is (−5,0), D is (−2,−2), E is (0,−3), and F is (3,−1).

We could also have said that B is (−3,+4) and that F is (+3,−1). Both are acceptable ways of writing coordinates.

It is essential to remember that:

(*a*) the *first* number is the x coordinate and goes *across* the page, to the right if positive, to the left if negative.

(*b*) the *second* number is the y coordinate and goes *up* the page if positive, *down* the page if negative.

Although we have not used fractions or decimals in any example so far, it is quite in order for us to do so. For example, the point halfway between (4,7) and (5,7) is the point $(4\frac{1}{2},7)$, or (4.5,7).

EXERCISE 6.7

1 What are the coordinates of the points I, J, K, L, M and N in the diagram?

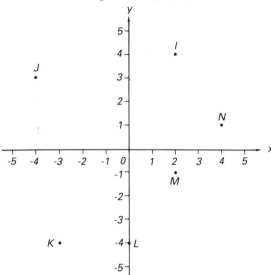

2 What are the coordinates of the points P and Q in the diagram?

(*a*) Write down the coordinates of four points which lie on the line joining P and Q.

(*b*) Does the point (2,1.5) lie on the line PQ? Write down some more points lying on PQ.

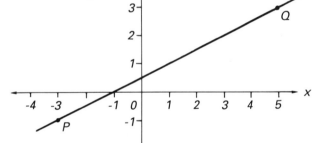

3 The diagram shows the first two steps in a set of seven going down.

(*a*) Copy the diagram and draw the steps on it.

(*b*) What are the coordinates of the two ends of step five? And of step seven?

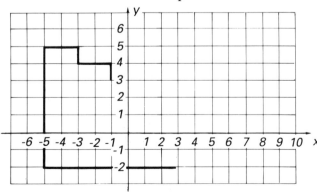

4 On a grid with axes from −5 to +5, plot the points V(−3,+5) and W(+3,−1).

(*a*) At what point does the line VW cut the *y* axis?

(*b*) Write down the coordinates of three other points on the line.

(*c*) Imagine that the line is extended in both directions. Write down the coordinates of any point which lies on the line to the left of V. Similarly, find a point on the line to the right of W.

5 Plot the points A(−3,−2), B(−3,4), C(−1,1), D(1,4) and E(1,−2) on a grid. Choose scales so that all the points fit on your diagram.

(*a*) What letter is formed by joining the points in order A, B, C, D, E?

(*b*) What letter is formed if C is connected with straight lines to the other four points?

6 Plot the points (4,5), (4,−1) and (−2,−1).

(*a*) If these three points are corners of a square, what are the coordinates of the fourth corner?

(*b*) What are the coordinates of the centre of the square?

(*c*) The diagonal from (−2,−1) to (4,5) cuts both axes. What are the coordinates of each of the points where this diagonal cuts the axes?

7 Join the points (−3,0), (−3,2) and (−3,−4) with a straight line.

(*a*) Write down the coordinates of two other points lying on this line.

(*b*) What do you notice about the *x* coordinate of each of these points? This line is called '*x*=−3', because all points on the line have an *x* coordinate of −3.

(*c*) On your diagram, draw the line *x*=−1, and write '*x*=−1' along it.

8 Join the points $(1,-5), (3,-5)$ and $(-4,-5)$ with a straight line.

   (*a*) Write down the coordinates of two other points lying on this line.

   (*b*) What do you notice about the *y* coordinate of each of these points? This line is called '$y=-5$', because all points on the line have a *y* coordinate of -5.

   (*c*) On your diagram, draw the line $y=-2$, and write '$y=-2$' along it.

9 (*a*) Write down the coordinates of the points A, B, C, D and E in the diagram.

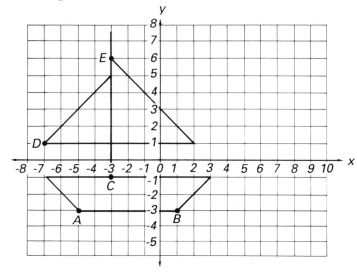

   (*b*) The 'ship' sails across the page, so that the point A moves to $(0,-1)$ and B moves to $(6,-1)$. By carefully plotting the points, draw the new position of the ship.

   (*c*) Write down the coordinates of the final positions of the points C, D and E.

   (*d*) What are the coordinates of the top of the mast after the ship has moved?

10 Draw a circle of radius 5 cm whose centre is at the origin $(0,0)$.

   (*a*) If you have been careful with your drawing, your circle will pass through the points $(4,3)$ and $(0,5)$.

   (*b*) Write down the coordinates of ten other points which lie on the circle.

7. Communication: graphics (2)

Drawing graphs from tables

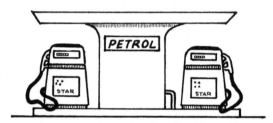

When a car is being filled up with petrol the display on the pump shows both the price and the number of litres of petrol increasing. When 1 litre has been put in the car, the price shows 38p, with 2 litres 76p, and so on. This kind of information can be represented by a graph.

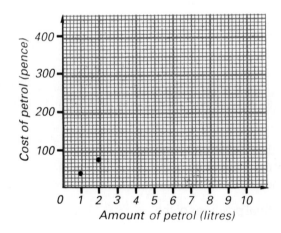

Copy the axes for the graph into your book. Work out how much 3 litres would cost, 4 litres, and so on. Plot these points on the graph as accurately as you can, and join them all with a straight line. We now have a graph we can easily use in the future without having to work things out again.

Use your graph to answer these questions:
(a) Find the cost of (i) 7 litres, (ii) $5\frac{1}{2}$ litres, (iii) $3\frac{1}{2}$ litres.
(b) If you only wanted to spend a certain amount of money, how much petrol would you get for (i) £1, (ii) £1.50, (iii) £1.25?

EXERCISE 7.1

Copy the axes into your book, and plot the points given in each table as accurately as you can. Answer the questions for each problem.

1 The speed of an accelerating car at 2 second intervals is shown in the table.

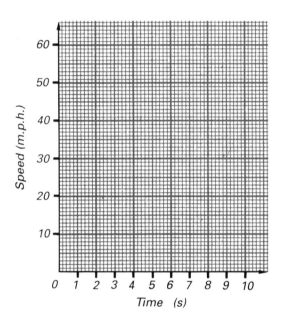

Time (seconds)	0	2	4	6	8	10
Speed (m.p.h.)	0	12	24	36	48	60

(*a*) Find the speed after (i) 5 seconds, (ii) $7\frac{1}{2}$ seconds, (iii) 3.2 seconds.

(*b*) How long does it take to get to (i) 15 m.p.h., (ii) 30 m.p.h., (iii) 57 m.p.h.

2 The cost of having invitation cards printed is shown in the table.

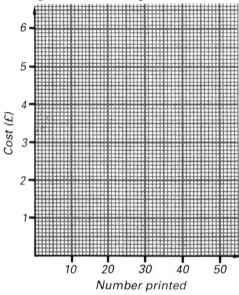

Number printed	10	20	30	40	50
Cost	£1.20	£2.40	£3.60	£4.80	£6.00

(*a*) Find the cost of having printed (i) 25 cards, (ii) 45 cards.

(*b*) How many cards would be printed for (i) £2, (ii) £5, (iii) £3.80?

3 The distance travelled by a car at a steady speed is shown in the table.

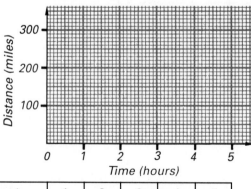

Time (hours)	1	2	3	4	5
Distance (miles)	60	120	180	240	360

(*a*) How far has the car travelled after (i) $2\frac{1}{2}$ hours, (ii) $\frac{1}{2}$ hour?

(*b*) After how long will the car have travelled (i) 90 miles, (ii) 255 miles?

4 The graph is to be used by a butcher to help calculate the cost of pork at various weights in kilograms.

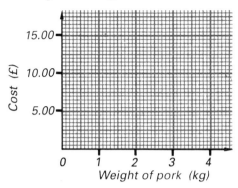

Weight of pork (kg)	1	2	3	4
Cost		£3.70	£7.40	£11.10	£14.80

(a) What weight of pork will cost (i) £5.50, (ii) £7.75?

(b) How much would these joints of pork cost: (i) 2½ kg, (ii) 1.75 kg?

5 The graph is to show the number of parts produced by an automated machine over a period of 50 minutes.

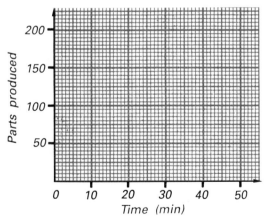

Time (min)	10	20	30	40	50
Parts	36	72	108	144	180

(a) How many parts would be produced after (i) 15 minutes, (ii) 35 minutes, (iii) 42 minutes?

(b) How many minutes would it take to produce (i) 90 parts, (ii) 100 parts, (iii) 150 parts?

Scaling

In all the graphs we have just completed we found it easy to plot the points because sensible scales had been chosen for the axes. When starting with a table of information we must choose the scales to be used. Look at the axes below and write down what you think is wrong with them. If in any doubt, try plotting the points in the tables; you may find it more difficult than before.

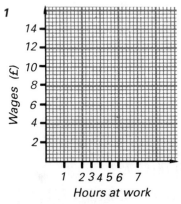

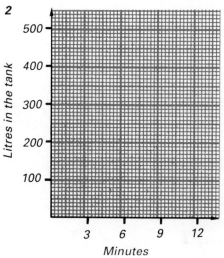

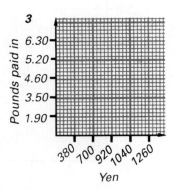

1 The amount earned over a number of hours at work is shown in the table.

Hours at work	1	2	3	4	5	6	7	8
Wages	£1.80	£3.60	£5.40	£7.20	£9.00	£10.80	£12.60	£14.40

2 The amount of water in a tank being filled after a number of minutes is shown in the table.

Minutes	2	4	6	8	10
Litres	100	200	300	400	500

3 Pounds converted into Japanese Yen is shown in the table.

Pounds	£1.90	£3.50	£4.60	£5.20	£6.30
Yen	380	700	920	1040	1260

EXERCISE 7.2

Redraw the axes the way you think they should be drawn for the tables and complete the graphs.

Straight lines or curves?

A graph sometimes represents a relationship between quantities shown on the axes. This could produce a straight line, or a curved graph.

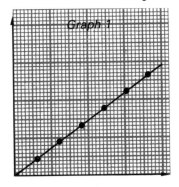
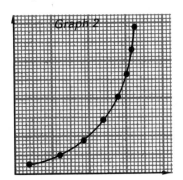
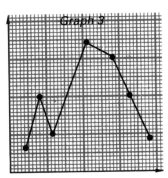

In graph 1 the points clearly lie on a line, and can be linked by a single straight line.

In graph 2 the points curve around, and should be joined by a curved line drawn *freehand*, as accurately as possible.

In graph 3 there appears to be no clear relationship between the points, so we join the points with individual straight lines in the order we plot the points.

EXERCISE 7.3

For each question draw a graph. Plan your scales carefully, plot the points and decide how you will join them up. Then answer the questions.

1 The number of hours of sunshine recorded during a ten-day period in summer is shown in the table.

Day	1	2	3	4	5	6	7	8	9	10
No. of hours	1	6	8	9	13	15	10	3	5	7

(*a*) What can be said about the hours of sunshine during the day?

(*b*) What could be said about the weather during this period?

2 The temperature of water as it is being heated is given in the table.

Time (s)	0	10	20	30	40	50
Temp. (°C)	10	27	44	61	78	95

(*a*) After how long was the temperature 50°C?

(*b*) What temperature was the water after (i) 5 seconds, (ii) 45 seconds?

3 Draw a graph to show the normal prices and the sale prices in a shop

Normal price (£)	1	2	3	4	5	6
Sale price (£)	0.85	1.70	2.55	3.40	4.25	5.10

(*a*) What can be said about the price reduction in the shop?

(*b*) What is the sale price of an article normally marked as (i) £1.60, (ii) £5.80?

4 The table shows the speed of a firework rocket as it climbs into the sky.

Time (s)	0	2	4	6	8
Speed (m/s)	0	60	100	130	155

(*a*) What is happening to the speed of the firework?

(*b*) What is the speed of the firework after (i) 3 seconds, (ii) 7 seconds?

(*c*) After how long is the speed (i) 80 m/s, (ii) 140 m/s?

5 The temperature taken every hour during a school day is recorded.

Time	0900	1000	1100	1200	1300	1400	1500	1600
Temp. (°C)	19	21	23	25	26	30	28	27

(*a*) Why do you think the temperature is higher around 2 o'clock?

(*b*) Could the temperature have risen above 30°C at any time during the day?

(*c*) By looking at the temperatures, can you deduce what time of year it is?

6 The table indicates the expected diameter of a tree as it grows.

Age (yrs)	10	20	30	40	50	60
Diameter (cm)	9	16	28	44	65	91

(*a*) What would be the diameter of a tree of age (i) 25 years, (ii) 48 years?

(*b*) When trees reach a diameter of 40 cm they can be cut down. After how long is this?

(*c*) Describe the way a tree grows from the information given in the graph.

EXERCISE 7.4

For each question draw a graph. Plan your scales carefully, plot the points and decide how you will join them up. Then answer the questions.

1 The table shows the distance of a cable car from its previous stop.

Time (min)	1	2	3	4	5	6
Distance (km)	1.2	2.4	3.6	4.8	6.0	7.2

(*a*) The cable car passes over a waterfall and a scree slope. How long does it take to travel from the previous stop to the waterfall and scree if their distances away from the previous stop are (i) 3 km, (ii) $5\frac{1}{2}$ km?

(*b*) How far has the cable car gone after (i) $3\frac{1}{2}$ minutes, (ii) $5\frac{1}{2}$ minutes?

2 The table indicates the speed of a car passing through the centre of a town as recorded at two-minute intervals.

Time (min)	0	2	4	6	8	10	12
Speed (m.p.h.)	30	20	0	15	0	25	35

(*a*) What can you say about the speed of the car?

(*b*) Give an explanation for the speeds taken after 4 and 8 minutes.

(*c*) What eventually happens to the speed of the car?

3 The monthly sales of a new brand of perfume from a large chemist chain is noted, and is to be presented in a graph.

Month	1	2	3	4	5	6
Sales	200	250	270	290	325	370

Give a description of the sales performance of the new perfume.

4 The table shows the weight of a foal as it grows.

Day	0	10	20	30	40	50	60
Weight (kg)	20	28	39	50	59	66	72

(*a*) Find the weight of the foal after (i) 25 days, (ii) 54 days.

(*b*) After how many days is the weight of the foal (i) 30 kg, (ii) 60 kg?

(*c*) Describe the way in which the foal increases in weight.

5 The information indicates the amount of petrol left in a tank as a car travels along a motorway at a constant speed.

Distance travelled (km)	0	20	40	60	80	100
Petrol left (litres)	36	33.5	31	28.5	26	23.5

(*a*) Find the amount of petrol left in the tank after (i) 24 km, (ii) 70 km.

(*b*) How far has the car travelled when there is just 25 litres in the tank?

(*c*) Give an explanation as to why you think the graph is the shape it is.

6 The time at which a train passed each $\frac{1}{4}$-mile post is shown in the table.

Time (s)	0	36	62	82	98	112	124	134
Distance (miles)	0	0.25	0.5	0.75	1	1.25	1.5	1.75

(*a*) How far had the train gone after (i) 45 seconds (ii) 2 minutes?

(*b*) A signal box is 0.6 mile from the station. How long did the train take to get there?

(*c*) What was happening to the speed of the train?

(*d*) What do you think will eventually happen to the speed of the train?

Investigation A

For this practical session you need a stop-watch and a piece of paper to note down the results of the experiment. Get a friend to run across the playground several times, and each time note down how long it takes them to complete a length. Using your results, plot a graph of the times of each run.

(*a*) What would you expect to happen?

(*b*) What did happen?

(*c*) What shape of graph did you get?

Extension

Compare the graphs of various people.

Investigation B

For this exercise you need a thermometer placed somewhere in the school. At regular intervals during the day make a note of the temperature, and use the information gained to plot a chart of the results.

(*a*) What shape was the graph?

(*b*) How would you explain to someone why the graph is this shape?

Extension

Plot a graph over several days for comparison.

Investigation C

For this practical you need access to a sink, a litre measuring jug and a stop-watch. Turn on the tap slowly, then leave it running. Time how long it takes you to measure 5 litres, 10 litres, 20 litres, and so on. Plot the results of the experiment on a graph.

(*a*) Were the results as you expected?

(*b*) What shape of graph did you get, and why do you think it is this shape?

Conversion graphs

Conversion graphs are a useful way to help us convert between one unit and another.

Example 1

Draw a conversion graph for inches and centimetres.

$$1 \text{ inch} = 2.54 \text{ cm}$$
so $\quad 2 \text{ inches} = 2 \times 2.54 = 5.08 \text{ cm}$
$$3 \text{ inches} = 3 \times 2.54 \text{ etc.}$$

We enter the answers in a table.

Inches	1	2	3	4	5	6	7
Centimetres	2.54	5.08					

Copy and complete the table. You can then use the table to plot the points on the scaled axes, joining them up to make the conversion graph.

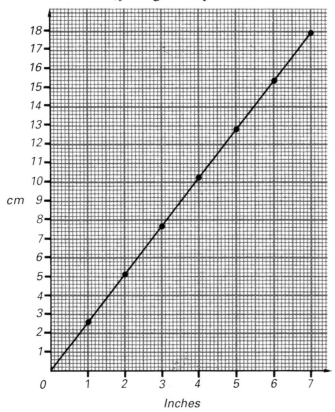

Use the graph to answer the following questions.

(*a*) Convert the following number of inches to centimetres: (i) 2.4 inches (ii) $4\frac{1}{2}$ inches (iii) 5.7 inches.

(*b*) Convert the following number of centimetres into inches: (i) 14 cm (ii) 6.4 cm (iii) 3.6 cm.

EXERCISE 7.5

For the following questions plot the points to make a conversion graph.

1

°F	40	50	60	70	80
°C	4.5	10	15.5	21	26.5

2

km/litre	10	12	14	16	18	20
m.p.g.	28.2	33.9	39.5	45.2	50.8	56.5

3

Pints	1	2	3	4	5	6	7
Litres	0.57	1.14	1.70	2.27	2.84	3.41	3.98

4 Conversion of French currency using a conversion rate of £1 = 9.80 francs.

£	1	2	3	4	5	6
French francs	9.80	19.60	29.40	39.20	49.00	58.80

EXERCISE 7.6

Now draw up tables yourself. Use the conversion given to work out 2×, 3×, 4×, etc., for your table, and draw the conversion graph.

1 £1 = 2.9 D-marks   **2** £1 = 1.10 I-punts

3 £1 = 230 yen   **4** 1 foot = 30.5 cm

5 1 ounce = 28.35 grams   **6** 1 cubic inch = 16.4 cubic cm

Investigation D

Many cartons, tins and packets of food items have their weight written on the side in both grams and ounces. Make a note of several of these, and plot a graph to compare grams with ounces. The information has provided you with a conversion graph. Select some other packets and check that your graph actually does work. How accurate is it?

Misleading graphs

Although graphs are used to make information clearer for us to understand, they can also be used to mislead people.

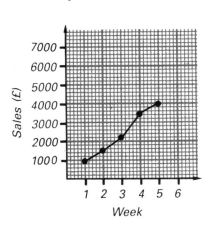 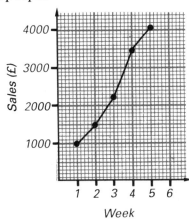

The two graphs represent the same information, but the second graph misleads as the incorrect scaling exaggerates the increase in the sales figures.

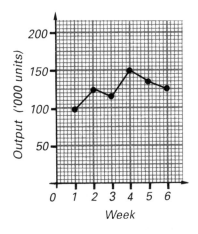 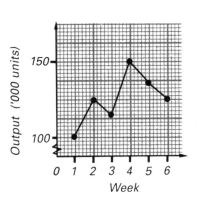

The two graphs again represent the same information. The elimination of the bottom portion of the second graph followed by the rescaling emphasizes the fluctuations of the graph.

EXERCISE 7.7

Explain why the graphs shown below can be used to deceive.

1 Redraw the graph correctly.

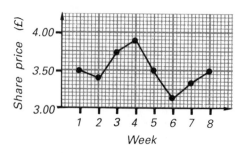

2 Redraw the graph correctly. **3**

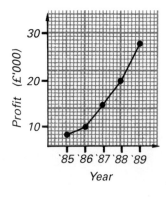 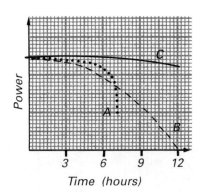

4 **5**

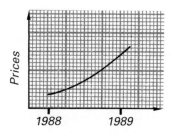 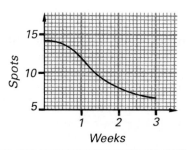

Investigation E

This is an extended project on shares. Start with £10 000, and the financial page of a newspaper. Decide which shares you will buy (shares must be bought in 100s), and prepare a portfolio – a list of shares you have bought, plus any money over. Every week check the prices of the shares you have bought. You can calculate percentage changes in prices, profit/losses, and draw a set of graphs to chart the value of your share issues.

Travel graphs

Travel graphs are a method of comparing the distance travelled over a period of time, but are also a useful way of representing a journey. The simplest travel graph is a single journey at constant speed. Normally speeds are simplified and averaged over short periods and as such the graph is only a mathematical *model* of the journey.

Example 2

A man goes out for a walk at 1300 hours, and walks for 2 hours at a steady rate of 3 m.p.h.

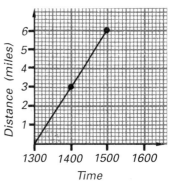

3 m.p.h. = 3 miles for every hour.
After 1 hour, distance covered = 3 miles.
After 2 hours, distance covered = 6 miles.

What does a small square represent on the horizontal time axis?

> 10 small squares = 1 hour *or* 60 minutes
> 1 small square = 60 ÷ 10 = 6 minutes.

Example 3

The graph shows a journey undertaken by cyclists. In the first 2 hours they covered 30 miles. Why do you think the graph is horizontal between 1100 and 1130 hours?

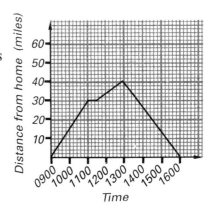

The horizontal portion of the graph means they were not moving. In this case they had stopped for lunch before starting again.

Why does the graph slope downwards at 1300 hours?

The cyclists had turned around and started on their way back home.

EXERCISE 7.8

1 A man goes out running for 2 hours at a speed of 8 m.p.h. Draw a simple one-line travel graph to fit this situation. How far is he from home after (*a*) $\frac{1}{2}$ hour, (*b*) 1 h 12 min?

2 A cyclist goes for a ride lasting 3 hours. His average speed is 15 m.p.h. Draw a simple travel graph.

(*a*) How far is he from home after $2\frac{1}{2}$ hours?

(*b*) How much time has elapsed after he has travelled $19\frac{1}{2}$ miles?

3 A moped rider travels at an average speed of 25 m.p.h., starting her journey at 11.00 a.m. and arriving at her destination at 3 p.m. Draw a simple travel graph for this situation.

(*a*) How far has she travelled after (i) $2\frac{1}{2}$ hours, (ii) 1 h 48 min?

(*b*) What time is it when she has travelled (i) 55 miles, (ii) 75 miles?

4 The graph shown is that of a jogger.

(*a*) How many miles have been covered after 3 hours?

(*b*) What is the average speed of the jogger?

(*c*) How long does it take the jogger to do 16 miles?

(*d*) How far has he gone after $2\frac{1}{2}$ hours?

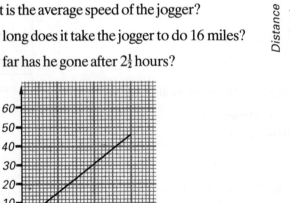

5

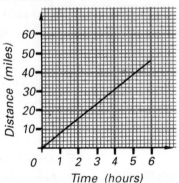

(*a*) How many miles has the cyclist gone after 6 hours?

(*b*) What is the average speed of the cyclist?

(*c*) How far has she gone in (i) 2 hours, (ii) $3\frac{1}{2}$ hours, (iii) 5 h 12 min?

(*d*) How long does it take her to cycle (i) 10 miles, (ii) 32 miles?

6 An aeroplane travels at 600 km/h on a journey lasting 5 hours. Draw a travel graph to represent this situation.

(*a*) How far has the aeroplane travelled after (i) $2\frac{1}{2}$ hours, (ii) 4 hours?

(*b*) How long has it been flying once it has gone (i) 2100 km, (ii) 2800 km?

EXERCISE 7.9

1 A car sets off on a journey at 8.30 a.m., and manages an average speed of 25 m.p.h. through the town, arriving at its destination at 1.00 p.m. Draw a travel graph to represent the situation.

(*a*) How far was the journey?

(*b*) How far had the car gone after (i) 3 hours, (ii) 4 hours?

(*c*) At what time did the car pass a service station 40 miles from its destination?

2 The graph represents a journey Sally made to see her friends.

(*a*) When did Sally first stop?

(*b*) For how long did she stop?

(*c*) What was the furthest distance she went from home?

(*d*) How long was the entire journey?

(*e*) When did Sally arrive home?

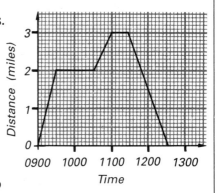

3 At 10.00 a.m. a salesman was taken by a colleague to pick up some display materials from Glasgow. He then flew back to Liverpool.

(*a*) How far from Liverpool is (i) Carlisle, (ii) Glasgow?

(*b*) For how long did they stop in Carlisle?

(*c*) When did the plane leave Glasgow?

(*d*) How long was the plane journey?

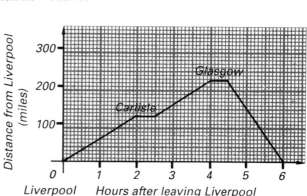

4 Mark sets off on a car rally at 11.30 a.m., making two stops on the outward journey. The car gives him some trouble on the return journey, and he has to stop at a garage. He then makes slower progress to the finish.

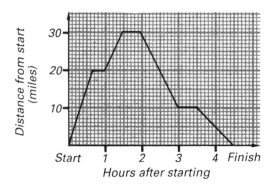

(*a*) How long was the first stop?

(*b*) How far was he from the start (i) at noon, (ii) at 2.00 p.m.?

(*c*) What time was it when he made (i) the first stop, (ii) the second stop?

(*d*) Which part of the journey was covered at the fastest speed?

5 Jacky leaves home at 1000 hours to walk to her friend's house, calling at the shops on the way. On arriving, she has to return home because she has forgotten something. On the way back to her friend's house she meets her Aunty, and stops to talk for a while.

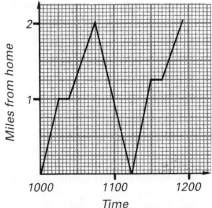

(*a*) How long were her stops?

(*b*) When did Jacky first arrive at her friend's house?

(*c*) How far from home was she when she met her Aunty, and at what time?

(*d*) How long did it take Jacky to return home?

(*e*) What was her average speed on the return journey home?

6 A salesman makes a number of calls during the day. Complete the time sheet below for his day's work.

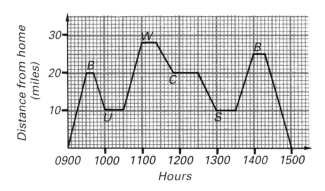

Place of visit	Time of arrival	Time of departure	Distance travelled from previous visit
Barton			
Uxbridge			
Witton			
Colehall			
Shilton			
Brompton			

Investigation F

Using the sames axes draw simple travel graphs for (*a*) a person walking at 3 m.p.h., (*b*) a person running at 10 m.p.h., (*c*) a cyclist travelling at 15 m.p.h. What do you notice about the slope of each graph? What do you think the slope represents?

Investigation G

Working in groups of 2 or 3, each of you draws a travel graph, then passes it to the next person in the group. Each person then has to write a different story to fit the travel graph. As an alternative you can start with a story and ask another person in your group to draw a graph to match the story.

Interpreting graphs

The graph shows the light given off by a fire. The fire gets bright very quickly once it has been lit, and stays bright for some time, then slowly becomes less bright until it is dim as the ashes glow.

A graph like this is a sketch to represent a situation and does not necessarily need accurate scaling. It conveys a message which must be understood: the shape of the graph is important.

EXERCISE 7.10

Each of the graphs shown represents the speed of a car driven a certain way. Match each of the descriptions given below with one of the graphs.

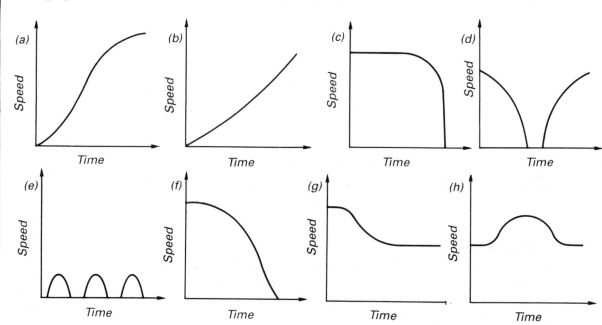

1 The car brakes slowly and stops.

2 The car comes off a fast motorway on to a slower road.

3 The car accelerates slowly.

4 The car is in a traffic jam.

5 The car is accelerating quickly.

6 The car does an emergency stop.

7 The car overtakes another car.

8 The car stops at traffic lights then starts again.

EXERCISE 7.11

Match the description of the wind speed for a day with that of the
graph of the wind speed.

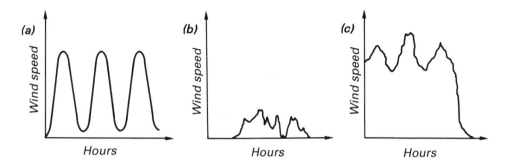

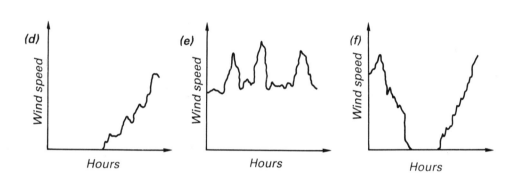

1 Windy all day, with occasional gusts.

2 Windy to start, with the wind dying down
around the middle of the day, and gaining
in speed again later.

3 Intermittent periods of calm with spells
of increased wind speed.

4 Very windy to start, but becoming
calmer later.

5 A few breezes during the middle of
the day.

6 Calm to start, getting windy towards
the end of the day.

EXERCISE 7.12

Match each of the situations below with one of the graphs shown.

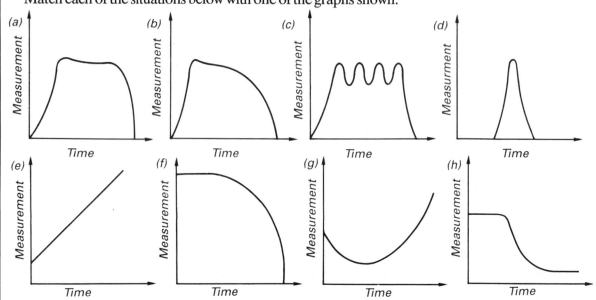

1 The temperature inside a kettle as it heats up from 40°C to 100°C.

2 After simmering, a pan of soup cools.

3 The police siren rises in pitch, warbles up and down while in use, then falls in pitch when turned off.

4 The central heating pump slowly builds up to speed, runs continuously while in use, then slows down gradually when turned off.

5 A train slows down to take a bend, then speeds up again.

6 A clockwork toy starts quickly once it is wound up, runs for a time, but begins to lose time as the spring winds down.

7 A measure of the light from a lighthouse as it turns around.

8 Your speed on a bicycle as you go up a steep hill.

EXERCISE 7.13

1 The following graphs describe the actions of a jogger. For each graph write an explanation of what you think might have happened.

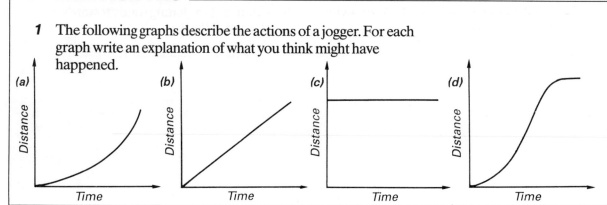

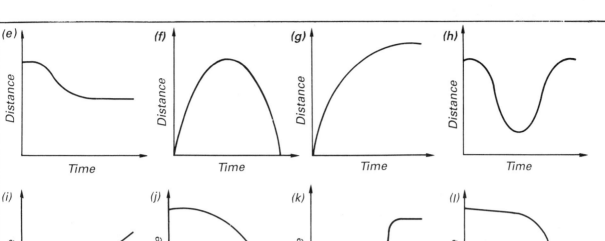

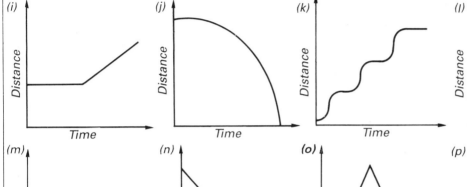

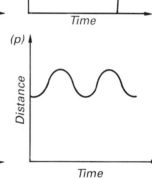

2 Draw a sketch graph for each of these situations:

(*a*) The amount of tape on a cassette spool as you play it through.

(*b*) How tired you feel as you go through the day.

(*c*) The total cost of renting a television if the first rental is due at the end of the month.

(*d*) The value of a new car over a number of years.

(*e*) The temperature during the day.

Investigation H

Complete the graph to show how busy you are at various times of the day.

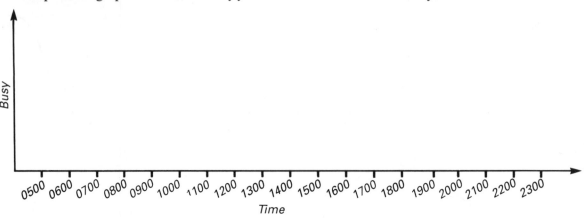

Investigation I

Complete the graph to show how quickly you are moving at various times of the day.

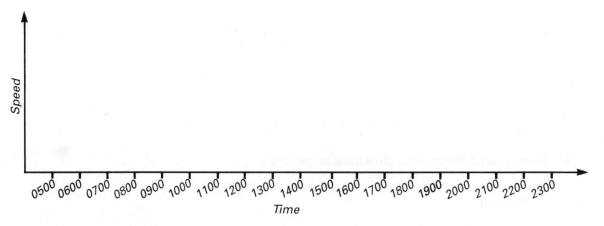

Using drawing instruments

Drawing instruments are essential for mathematics, particularly when we need to construct accurate diagrams. The instruments you need are shown in the diagram.

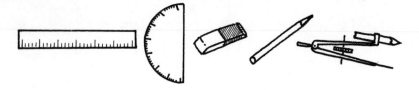

Rectangular drawings

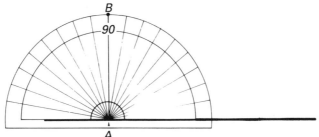

To draw a 90° angle at point A you should position the protractor as shown in the diagram, mark point B at 90°, then draw in the line from A to B. Check to make sure it is exactly 90°. Each time you need a right angle in an accurate drawing you should do this.

EXERCISE 7.14

Draw the following shapes accurately, using a protractor and a ruler.

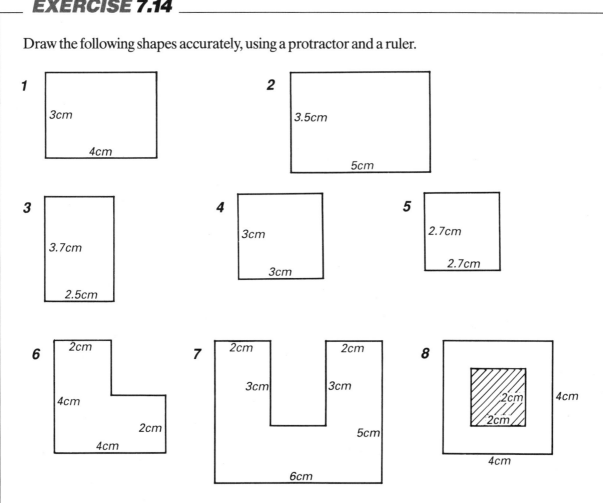

1   3cm   4cm

2   3.5cm   5cm

3   3.7cm   2.5cm

4   3cm   3cm

5   2.7cm   2.7cm

6   2cm   4cm   2cm   4cm

7   2cm   2cm   3cm   3cm   5cm   6cm

8   2cm   2cm   4cm   4cm

Triangles

Example 4

Draw the triangle ABC accurately.

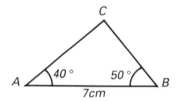

Draw the line AB first, making it 7 cm long. Use your protractor to draw angle A = 40°, then angle B = 50°. Extend the lines from A and B to meet at point C. You then have the complete triangle.

EXERCISE 7.15

Draw these triangles accurately.

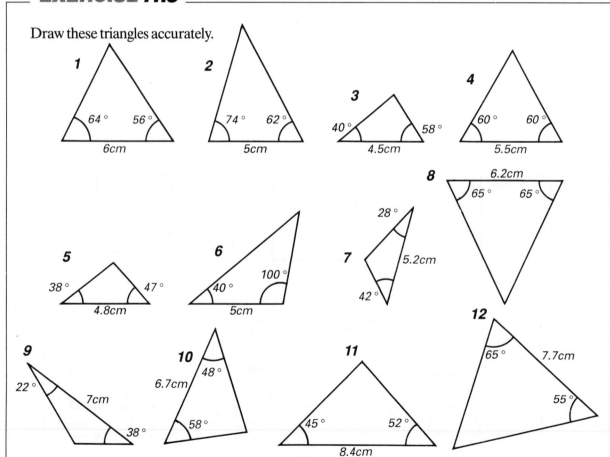

Example 5

Draw the triangle ABC accurately.

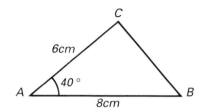

Draw the line AB first, making it 8 cm long. Draw angle A = 40° using your protractor, and draw in line AC, making it 6 cm long. Join C to B, and the triangle is complete.

EXERCISE 7.16

Draw these triangles accurately:

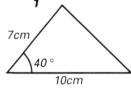

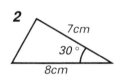

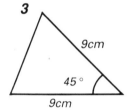

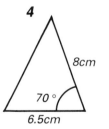

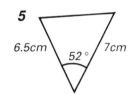

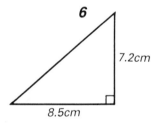

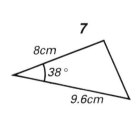

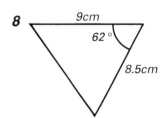

Example 6

Draw the triangle ABC accurately.

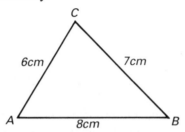

You now need your pair of compasses. Start by drawing line AB to be
8 cm long. Set your compasses to 6 cm, and with the point on A make a
mark on the page as shown below. Similarly set your compasses to 7 cm,
put the point on B and make a mark on the page as before.

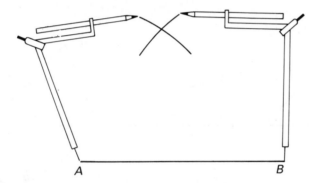

When you have made the cross on your page, label it C. Join A to C, then
B to C, and the triangle is complete.

EXERCISE 7.17

Draw the triangles accurately.

1 — 7cm, 9cm, 8cm

2 — 3cm, 4cm, 5cm

3 — 7.9cm, 8.5cm, 7.3cm

4 — 6.3cm, 5.4cm, 7cm

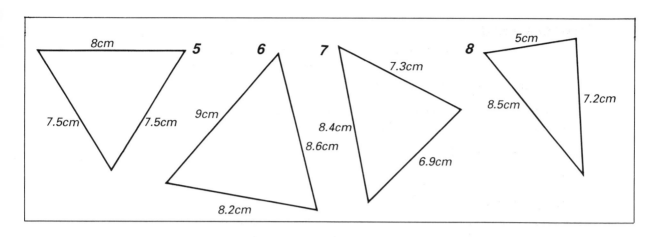

EXERCISE 7.18

Draw these shapes accurately.

1

3cm   5cm

4.5cm

2

70°

6cm

6cm

3

40°   40°

5.5cm   5.5cm

6.5cm

4

7cm   7cm

68°   68°

10cm

5

50°   8cm   50°

50°   50°

6

5cm 40°   40° 5cm

7.5cm

7

4cm

4cm 120°   120° 4cm

120°   120°

120°   120° 4cm

4cm

8

120°

All lengths 3cm

Investigation J

The following instructions will help you to draw accurately a series of polygons. For each regular polygon use a side of length 3 cm. The interior angle you will have to use for each polygon is given below.

square: 90°   pentagon: 108°   hexagon: 120°
octagon: 135°   decagon: 144°

Add up the angles of each of these polygons. Do you notice anything about them?

Investigation K

Work in pairs for this exercise. One of you sketches a triangle and puts some information on it, like lengths of sides, angles, etc., but not the same each time! The other one then has the job of accurately drawing the triangle. Repeat the exercise for several different triangles.

(*a*) What do you notice about the triangles you can't draw?

(*b*) What is the minimum amount of information you need to be able to draw a triangle?

Note: Be practical! Keep angles less than 90° and measurements less than 10 cm.

8. Communication: geometric

Triangles and quadrilaterals

Many simple shapes that we see can be thought of as being made up of a
number of *triangles* – the simplest straight-edged shape which is enclosed.
  Two lines can't form a closed shape, but *three* lines can, giving a **triangle**.

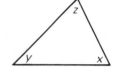

(*a*) Draw a triangle on card, or thick paper, and cut it out. Any triangle
will do.

(*b*) Now put your triangle on your page (or a separate piece of paper)
and draw round it.

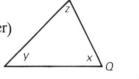

(*c*) Rotate your triangle, so that X touches P and Y touches Q. Draw
round the new position.

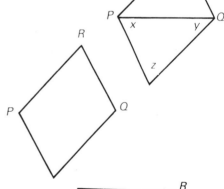

(*d*) You now have a four-sided shape, called a **quadrilateral**.

(*e*) Now slide your original triangle so that ZY is along PR. Draw round
it.

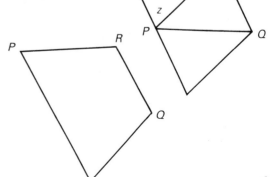

(*f*) You now have another quadrilateral.

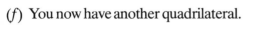

143

Investigation A

Continue repeating rotations and slidings, each time increasing your pattern, to obtain more complicated shapes.

Investigation B

Repeat the process above with a different triangle.

Extension

(*a*) Choose a triangle with a square corner (i.e. a right angle, or 90°) and repeat the process above.

(*b*) Use an equilateral triangle; you should get a regular (boring!) pattern.

Investigation C

Can you cover the whole of your sheet of paper by starting with *any* triangle?

What about starting with any quadrilateral?

Let us look more closely at triangles.

Measure the angles in this triangle, with a protractor.

You should find that angle D = 70°, angle E = 35°, and angle F = 75°.
Now add these up: the total is 180°.

Draw any triangle you like on your paper. Measure the angles and add them up.

Do the same, with another triangle.

You should find that your total is close to 180°. If not, check your angles again.

It can be shown that

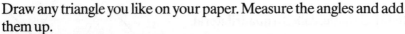

the sum of the angles in any triangle = 180°

Draw a line 9 cm long. Draw an angle of 54° at one end, and an angle of 65° at the other.

Measure the angle at the top, at M. Can you calculate what angle M should be?

As all three angles have to add up to 180°, we can say that

$$54° + 65° + M = 180°$$
$$119° + M = 180°$$
So
$$M = 180° - 119°$$
$$= 61°$$

How near were you?

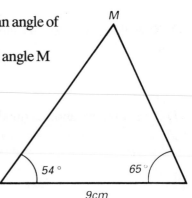

EXERCISE 8.1

In questions 1–6, work out the angles marked by small letters.

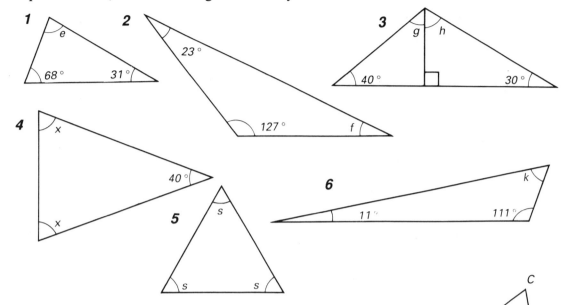

7 Draw a line AB, 11.2 cm long. Draw an angle of 37° at A and an angle of 79° at B, and complete the triangle ABC. Measure the angle at C.

Now *calculate* what angle C should be.

8 Draw triangle PQR, with PQ = 12.7 cm, angle P = 80°, and angle Q = 26°. Calculate angle R, and measure it in your diagram.

9 Draw a line XY, 6.8 cm long, starting near the left-hand side of your page. At X draw an angle of 30°, and at Y draw an angle of 120°. (Take care – this angle is greater than a right angle.) Complete the triangle, calling the top point Z. Measure the angle at Z. What should it be?

By considering angles X and Z, how long do you think line YZ is? Measure it. Were you right?

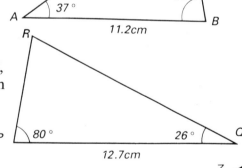

10 In the diagram, angle T = 90°, RT = 6 cm and TS = 8 cm.

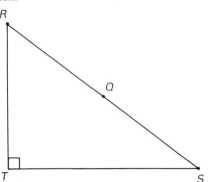

(*a*) Draw the triangle RST.

(*b*) Measure angles R and S.

(*c*) How long is side RS?

(*d*) Find the midpoint of RS. Call it Q.

(*e*) Join Q to T.

(*f*) Measure the distance QT.

(*g*) Measure angles QTR and QTS.

(*h*) What name do we give to triangles like RTQ and QTS, which have two equal angles?

Investigation D

(*a*) Draw a line AB, 12 cm long.

(*b*) Draw a line AX through A, sloping up at about 30°.

(*c*) Draw a line from B at right angles to AX, marking the point where it meets AX with a dot.

(*d*) Repeat steps (*b*) and (*c*), but with different angles, from 10° to 80°. What shape do the dots make?

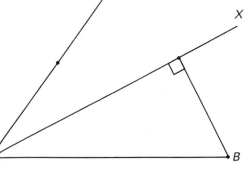

Extension

(*e*) Draw a line through A sloping *down* at various angles.

(*f*) Repeat Investigation D using an angle of 80° instead of 90°.

(*g*) Try other angles.

From earlier work, you should remember that the angles on a straight line add up to 180°. (From one end of a line, swinging round to the other, you complete a half-turn, or half of 360°, which is 180°.)

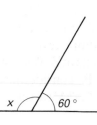

So in the diagram, angle $x = 180° - 60° = 120°$.

   Using this information we can work out angles on rather more complicated diagrams.

Example

Work out angles a and b in this triangle.

Angle $a = 180° - 111°$
$\qquad = 69°$

As $a + b + 64° = 180°$,

$\qquad 69° + b + 64° = 180°$
$\qquad\quad b + 133° = 180°$
$\qquad\qquad\qquad b = 180° - 133°$
$\qquad\qquad\qquad\quad = 47°$

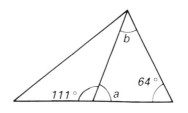

EXERCISE 8.2

Work out each of the lettered angles.

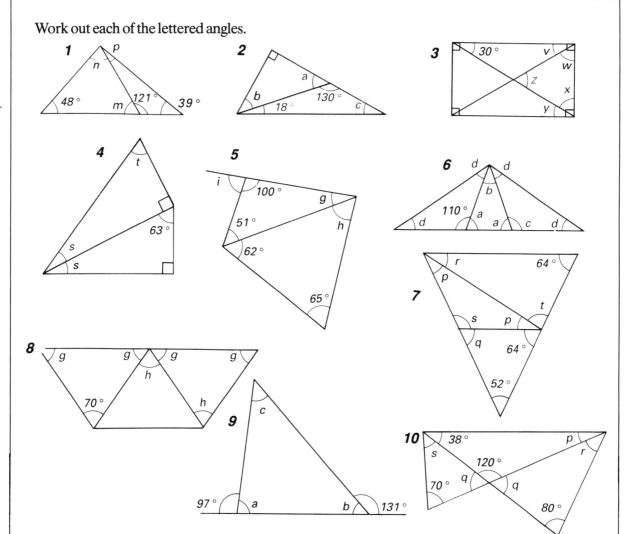

Investigation E

Draw any quadrilateral. Draw a diagonal. Can you draw another diagonal?

Can you find any quadrilaterals in which you can draw only *one* diagonal?

What distinguishes these quadrilaterals from those in which you can draw *two* diagonals?

We can draw a diagonal across any quadrilateral.

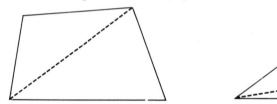

In *each* of the *two* triangles formed the angles total 180°. So in any quadrilateral, which is made up of two triangles, we can say

the sum of the angles of a quadrilateral is 360°

EXERCISE 8.3

Work out the angles marked by letters in these diagrams.

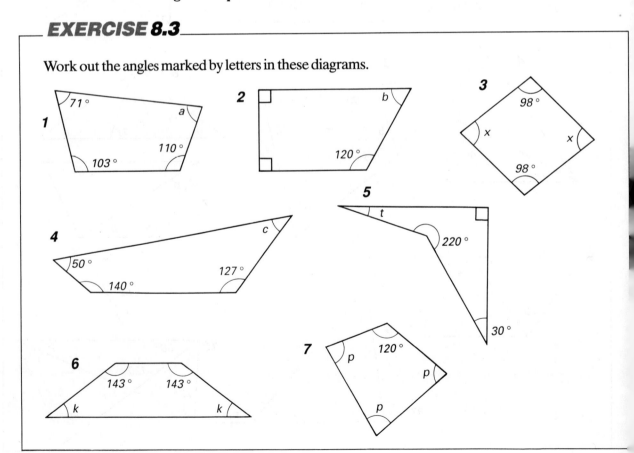

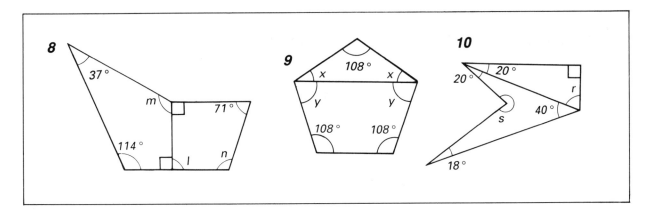

It is helpful to have a list of some of the more common words used in
conjunction with straight-sided shapes.

Triangle – a figure with three straight sides.

Quadrilateral – a figure with four straight sides.

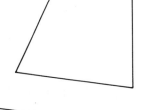

Polygon – a figure with any number of straight sides.

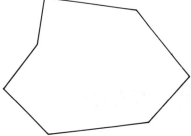

Diagonal – a line joining two corners of a polygon, and crossing the
inside of the polygon.

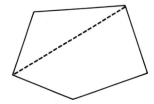

Degree – a measure of angle. There are 360° in one full turn.

Right angle – a quarter of a turn, or 90° (degrees).

Acute angle – less than a right angle (from 0°–90°).

Obtuse angle – more than a right angle but less than two right angles (from 90°–180°).

Parallel lines – lines pointing in the same direction, always the same distance apart.

Perpendicular lines – lines at right angles to each other (they *don't* have to be vertical or horizontal).

Circles

Use your compass to draw some circles. The distance between your pencil point and the compass point is the *radius* of the circle.

You can make some interesting patterns with circles. Here are some ideas for you to try.

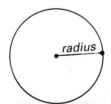

radius

Investigation F

(*a*) (i) Draw a circle (about 5 cm – 7 cm radius).

 (ii) Put the compass point anywhere *on* the circle and draw a curve across the inside of the circle, without altering the radius of your compass. Your curve should pass through the centre of the circle. Make it touch the edge of the circle at both ends.

 (iii) Now put the compass point on one end of the curve, and make another curve like the first one, across the inside of the circle.

 (iv) Repeat, until you get the final pattern.

 (v) Now colour it in.

(*b*) (i) Draw three equal circles, each touching the other two.

 (ii) Try to draw a circle in the gap in the middle, so that it *just* touches all three circles.

 (iii) Try to draw a circle which *just* encloses all three circles.

(*c*) What sort of triangle do you get if you join the centres of three equal circles which touch each other?

Here is a list of common words used when talking about circles.

Radius – distance from the centre of the circle to any point on the circle.

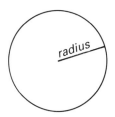

Diameter – distance across the circle, through the centre. This distance is always equal to *twice* the radius. (A penny has a diameter of 2 cm – check this.)

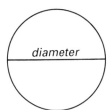

Circumference – distance round the outside of the circle. It is a little over *three* times the diameter.

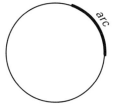

Arc – a part of the circumference.

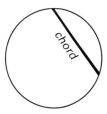

Chord – a line joining any two points on the circumference.

9. Communication: transformations

Scale drawings

This rectangle is drawn to a scale of 1 cm to 2 cm, which means that each 1 cm in the diagram represents an actual length of 2 cm.
   What is the actual length? 3 cm × 2 = 6 cm
What is the actual width? 2 cm × 2 = 4 cm
Draw the scale diagram, and the actual rectangle it represents.

This triangle is drawn to a scale of 1 cm to 3 cm.
What is the actual height? 2.5 cm × 3 = 7.5 cm
What is the actual width? 1.5 cm × 3 = 4.5 cm
Draw the scale diagram, and the actual triangle it represents.

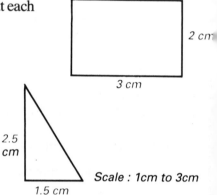

Scale : 1cm to 2 cm

2 cm

3 cm

2.5 cm

1.5 cm

Scale : 1cm to 3cm

EXERCISE 9.1

Redraw each diagram to its actual dimensions.

1
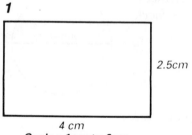
2.5cm
4 cm
Scale : 1cm to 2cm

2
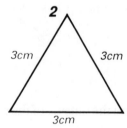
3cm   3cm
3cm
Scale : 1cm to 2 cm

3
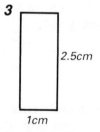
2.5cm
1cm
Scale : 1cm to 3cm

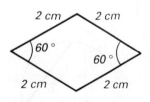
2 cm   2 cm
60°
60°
2 cm   2 cm
Scale : 1cm to 2.5cm

5
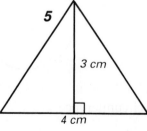
3 cm
4 cm
Scale : 1cm to 1.5cm

6
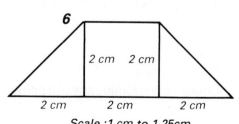
2 cm   2 cm
2 cm   2 cm   2 cm
Scale :1 cm to 1.25cm

EXERCISE 9.2

In questions 1–5, copy the diagram, but write in the actual dimensions of the shape to the given scale.

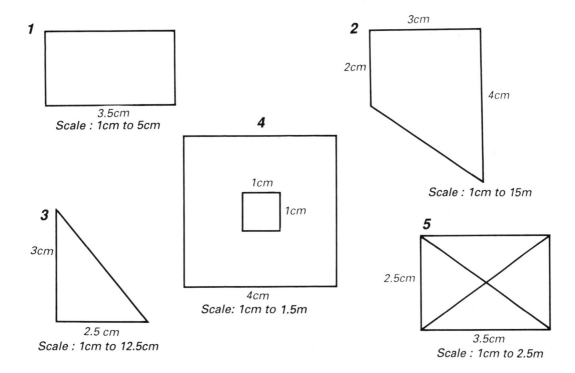

1

3.5cm
Scale : 1cm to 5cm

2

3cm

2cm

4cm

Scale : 1cm to 15m

4

1cm

1cm

4cm
Scale: 1cm to 1.5m

3

3cm

2.5 cm
Scale : 1cm to 12.5cm

5

2.5cm

3.5cm
Scale : 1cm to 2.5m

6 (*a*) Draw an exact copy of the shape.

(*b*) Calculate the actual lengths of the sides to the given scale and write them on your diagram.

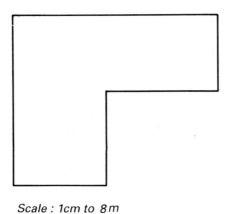

Scale : 1cm to 8m

7 (*a*) Draw an exact copy of the shape.

(*b*) Calculate the actual lengths of the sides to the given scale and write them on your diagram.

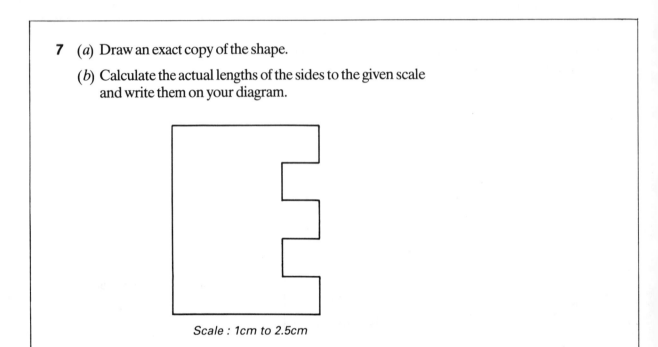

Scale : 1cm to 2.5cm

You cannot draw this diagram accurately in your book as it is too large. A scale diagram is needed.

A scale of 1 cm to 2 cm would reduce the actual length from 30 cm to 15 cm.

A scale of 1 cm to 3 cm would reduce the actual length from 30 cm to 10 cm.

A scale of 1 cm to 4 cm would reduce the actual length from 30 cm to 7.5 cm.

A scale of 1 cm to 5 cm would reduce the actual length from 30 cm to 6 cm.

30cm

30cm

Which scale would you use to draw the scale diagram? You should choose a scale so that the diagram fits on the page easily. Pick a scale and draw a scale diagram of the square. Don't forget to write down in your book the scale you have used near to the diagram.

Example 1

The diagram represents a small plot of land. Copy and complete the table below to find the measurements you would use for the length and width of the rectangle for each of the scales shown.

48m

18m

Scale	Length	Width
1 cm to 2 m		
1 cm to 3 m		
1 cm to 4 m		
1 cm to 5 m		
1 cm to 6 m		

Choose one of the scales and draw a scale diagram of the plot of land.
Also write down next to the diagram the scale you have used.

EXERCISE 9.3

For each problem you should choose an appropriate scale, draw a
scale diagram and write the original measurements on the diagram.
Remember to write down next to each diagram the scale you have
used.

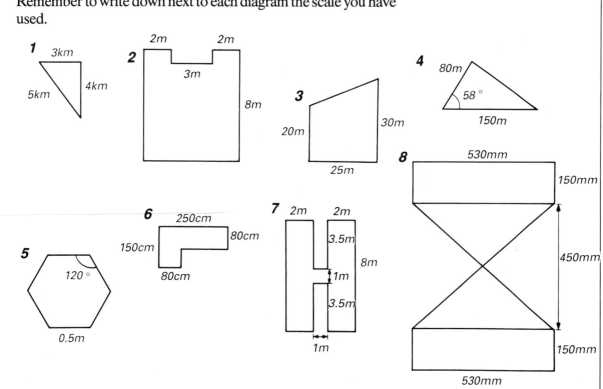

EXERCISE 9.4

1 The diagram shows a plan of the ground floor of a house, drawn to scale. For each room find (*a*) the length, (*b*) the width, (*c*) the area.

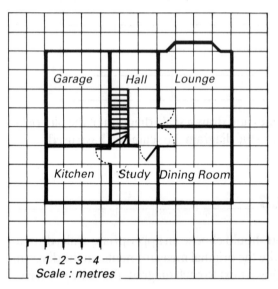

2 The diagram shows a plan of the first floor of a house, drawn to scale. For each room find (*a*) the length, (*b*) the width, (*c*) the area. Compare the plan with the one in question 1. Why is the plan much smaller?

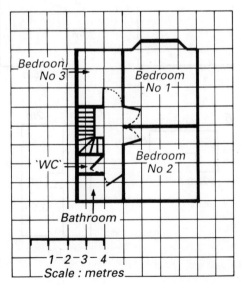

3   The diagram shows a plan of the floor of an empty office, drawn
to scale.

1   2   3   4   5
Scale : metres

(*a*) Find the total floor space inside the office.

(*b*) Desks are to be put in the office, but there must be at least a
space of 2 m × 2 m for each desk and office worker. How
many desks can be put into the office?

(*c*) Telephone sockets are to be installed along each wall,
spaced out at 4 m intervals along the base of each wall, and
at least 5 m from a corner of the office. How many will be
fitted?

Investigation A

Produce a scale drawing of the classroom, including the position of the
blackboard, and any doors and windows.

Investigation B

Martin has the job of stacking boxes into a rack as shown in the diagram.

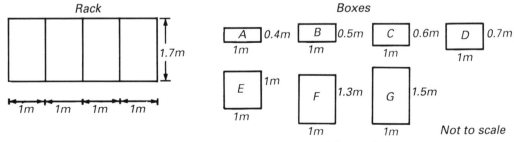

Make a scale drawing of the rack, and find out how you might stack the
boxes so that they fit neatly into the four parts of the rack. Complete your
scale drawing by adding the boxes to your diagram.

Investigation C

Plan out a kitchen on squared paper. Start with a scale drawing on squared paper and add units of the right size to fit. Keep a list of all the units you have used, and the cost. The kitchen has dimensions 3.3 m × 2.9 m with a door in the middle of each of the shorter walls. You will also need to fit in the following items, whose widths are shown:

cooker 55 cm, fridge 50 cm, washer 60 cm, sink unit 100 cm

You will need to use these details of kitchen units available from a retailer to find the dimensions of each unit and the total cost.

Full Door Base Units	300mm	500mm	600mm	1000mm	1200mm	1100mm Corner using 500mm Door	1100mm Corner using 600mm Door
Co.	37.80	38.80	41.30	54.75	61.75	56.90	59.40
Tr.	47.80	48.80	52.80	74.75	84.75	66.90	70.90
Pi.	47.80	48.80	52.80	74.75	84.75	66.90	70.90
We.	48.80	49.80	54.40	76.75	87.95	67.90	72.50

Door and Drawer Base Units	500mm	600mm	1000mm	1200mm	1100mm Corner using 500mm Door	1100mm Corner using 600mm Door
Co.	51.70	54.80	80.55	88.75	69.80	72.90
Tr.	65.30	69.70	107.75	118.55	83.40	87.80
Pi.	65.30	69.70	107.75	118.55	83.40	87.80
We.	66.70	71.90	110.55	122.95	84.40	90.00

720mm high Wall Units	300mm	500mm	600mm	1000mm	1200	600mm Corner	600 × 350mm Hob Wall
Co.	36.80	36.80	39.30	55.75	61.75	41.80	44.30
Tr.	46.80	46.80	50.80	75.75	84.75	51.80	54.30
Pi.	46.80	46.80	50.80	75.75	84.75	51.80	54.30
We.	47.80	47.80	52.40	77.75	87.95	52.80	55.30

4 Drawer Unit	500mm	600mm	3 Drawer Unit 600mm	2 Drawer Unit 600mm
Co.	72.45	74.85	87.30	99.75
Tr.	86.85	88.45	104.10	119.75
Pi.	86.85	88.45	104.10	119.75
We.	88.45	90.86	106.30	121.75

Larder Unit	500mm	600mm	Single Oven Unit 600mm	Double Oven Unit 600mm
Co.	115.85	123.35	111.45	127.90
Tr.	145.85	157.85	137.85	147.90
Pi.	145.85	157.85	137.85	147.90
We.	148.85	162.65	141.65	149.90

720mm high Glass Wall Unit	500mm	1000mm	720mm high Open End Wall Unit	Open End Base Unit 600mm
Co.	65.40	112.95	Honey 34.95	Honey 44.95
Tr.	82.40	146.95		
Pi.	82.40	146.95	Pine 34.95	Pine 44.95
We.	83.40	148.95		

Under Oven Base Unit	600mm
Co.	28.40
Tr.	31.80
Pi.	31.80
We.	32.80

Decorative End Panels
Wall Unit 5.59
Oven/Larder Unit 31.18 (pkt of 2)
Base Unit 9.30
4 Drawer Pack 14.65 *(please state) LH or RH)
Handles Pine Knobs (pkt of 10) 6.50
Westminster Door (pkt of 10) 10.95
Drawer (pkt of 2) 2.75
Traditional Door (pkt of 8) 7.40
Drawer (pkt of 2) 1.95

N.B. Base units are 900mm high with worktop fitted.
Wall units are 720mm high.
Hob wall unit is 350mm high.
Larder and oven units are 2130mm high.
All drawer base prices include appropriate drawer box.
Co. = Country   Tr. = Traditional   Pi. = Pine   We. = Westminster

Investigation D

Draw the scale diagram of the plan of the bedroom. Using the same scale, draw and cut out plans for the following pieces of furniture:

bed	190 cm by 90 cm	wardrobe	80 cm by 50 cm
pouffe	35 cm diameter	bedside cupboard	30 cm by 40 cm
chair	40 cm by 40 cm	chest of drawers	60 cm by 50 cm
dressing table	100 cm by 35 cm		

By moving around the pieces of furniture on the plan, decide where you want each piece of furniture. Draw on to the plan your final positions for the furniture.

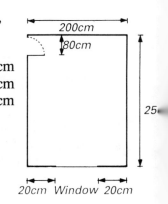

200cm
80cm
25
20cm Window 20cm

Enlargement

Enlargement is a term used frequently in photography when a photograph is made larger.

(a)

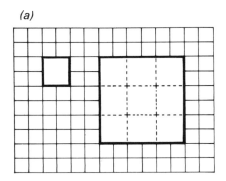

(b)

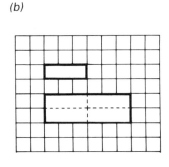

(c)

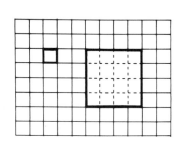

How much longer have the sides of the shape been made in (a), (b) and (c)?

Shape (a) has sides 3 times longer, shape (b) twice as long, and (c) four times as long. This is what we call the scale factor of the enlargement.

Example 2

If a rectangle of dimensions 3 cm × 4 cm is enlarged by a scale factor 2 its dimensions become 6 cm × 8 cm. An enlargement by a scale factor 3 makes it 9 cm × 12 cm.

EXERCISE 9.5

For each question, copy the diagram as given, and then draw the enlargement to the given scale factor.

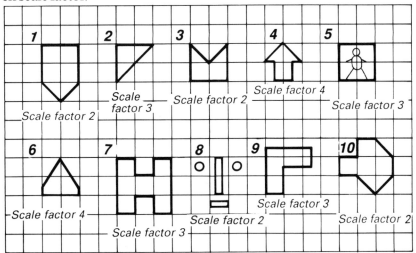

As well as having a scale factor, an enlargement should also have a point of enlargement.

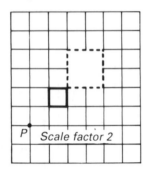 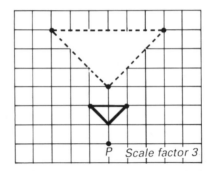

In each of the enlargements, P is the point of enlargement. Not only is the shape enlarged, but the distance of each corner from the point P is also increased by the same scale factor.

Alternatively, if plotted on a pair of axes, the same would apply.

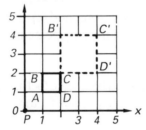

A(1,1)→A′(2,2); B(1,2)→B′(2,4); C(2,2)→C′(4,4); D(2,1)→D′(4,2)

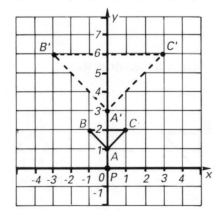

A(0,1)→A′(0,3); B(1,2)→B′(3,6); C(1,2)→C′(3,6)

Do you notice anything about the coordinates? By how much has each coordinate been increased? By the same factor as the scale factor. As long as the point of enlargement is at the origin this is an alternative way of working out the position of the enlarged figure.

EXERCISE 9.6

For each question copy the diagram as given and then draw the enlargement to the given scale factor from the point of enlargement.

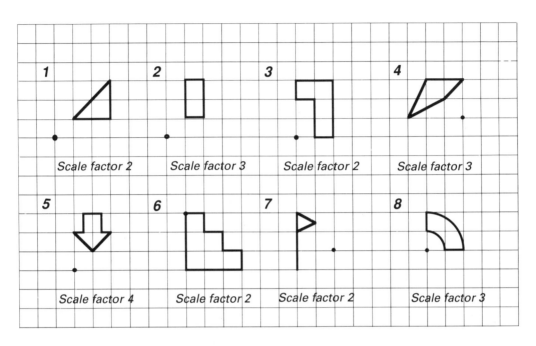

1 Scale factor 2

2 Scale factor 3

3 Scale factor 2

4 Scale factor 3

5 Scale factor 4

6 Scale factor 2

7 Scale factor 2

8 Scale factor 3

9 Copy the shape, and on the same diagram produce enlargements of scale factor 2 about the points of enlargement A, B, C and D.

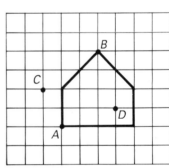

EXERCISE 9.7

For each question draw a pair of axes, plot the coordinates and join them up to give a shape. Then draw the enlargement to the given scale factor, using (0,0) as the point of enlargement.

1 A(1,2), B(1,3), C(2,3), D(2,2), scale factor 3.

2 A(0,3), B(3,0), C(3,−2), D(0,1) scale factor 3.

3 A(−2,3), B(4,3), C(4,−2), D(−2,−2), scale factor 2.

4 A(−2,3), B(2,3), C(2,−2), scale factor 3.

5 A(1,3), B(4,3), C(4,2), D(3,2), E(3,−1), F(2,−1), G(2,2), H(1,2), scale factor 2.

6 A(−2,3), B(2,3), C(2,2), D(1,1), E(1,0), F(−1,0), G(−1,−2), H(−2,−2), scale factor 2.

Investigation E

In the above exercises we placed the point of enlargement at the origin of coordinates. What would happen if we placed it somewhere else? Place the point of enlargement near to, but not at the origin, and repeat the problems given. Does this change the shape of the enlargement? Is there any pattern to the coordinates? Repeat with a few examples of your own if you are unsure.

Investigation F

Select a small object to draw around on squared paper, and attempt to produce an enlargement of it. State clearly the scale factor you have used.

Investigation G

Enlarge a drawing in a book by a factor of 2.

  Take a sheet of tracing paper the size of the drawing, and a sheet of drawing paper twice the dimensions of the tracing paper. Draw centimetre squares lightly in pencil on both sheets, and use the tracing paper to make a copy of the drawing. You should now be able to produce an enlargement of the drawing on to the paper using the squares to help you.

Line symmetry

In Book 1, Chapter 13, you were shown how to draw shapes which are
symmetrical about a line.

   Pin prick through the corners of the shape, open it out and draw in the
opposite side. The crease in the paper is called the line of symmetry,
and the shape formed is said to have line symmetry about this line.

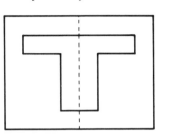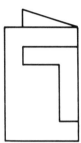

EXERCISE 9.8

For each question, copy and complete the diagram, drawing in the
reflection of the shape in the line of symmetry.

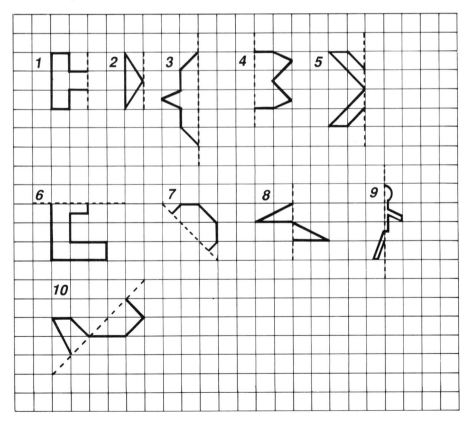

EXERCISE 9.9

Copy the shape and draw in a line of symmetry.

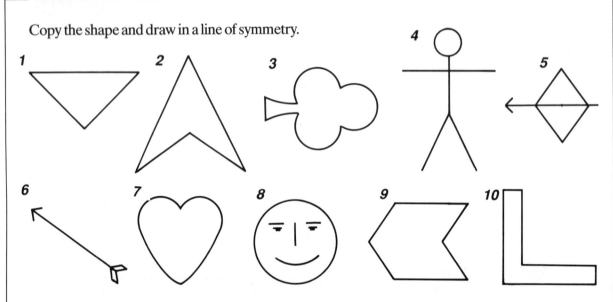

1
2
3
4
5
6
7
8
9
10

Some shapes have more than one line of symmetry, others may have none. How many does a square have?

Copy the square on to a sheet of paper, cut it out and find out how many different ways you can fold the shape on to itself. Do you get 4 ways? The square has 4 lines of symmetry.

Is this a line of symmetry? If we were to fold the shape over using the dashed line as a crease, we would find that it does not match exactly.

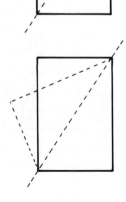

How many lines of symmetry does a rectangle have?

EXERCISE 9.10

For each question, copy the diagram and draw in all the possible lines of symmetry. Write down how many you have found in each case. You may find some of the shapes do not have any lines of symmetry.

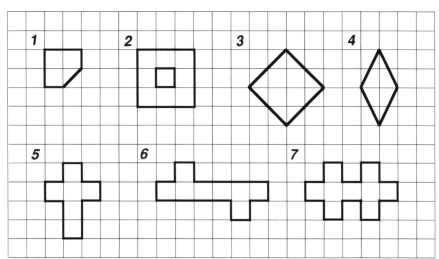

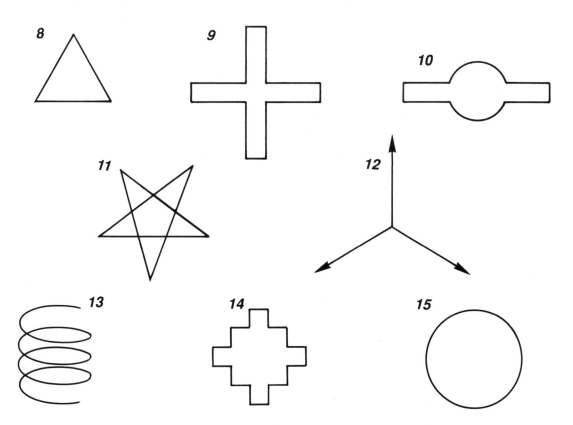

Investigation H

Draw an object which you know is symmetrical, and find out how many lines of symmetry it has. Try to find a few objects with more than one line of symmetry, and draw them in your book.

Investigation I

The method described in the book for producing diagrams with one line of symmetry involved using a pin. Can you devise a method for producing patterns which have 4, or 8, lines of symmetry using the same method? You might have to fold the piece of paper different ways.

Investigation J

Some words such as 'bid' are themselves symmetrical. How many other words do you know like this?

Reflection

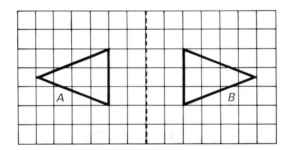

Triangle A has been reflected in the line to form triangle B. This is a similar idea to line symmetry, but one shape creates a second.

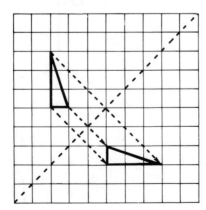

Each point of an **object** is reflected in the line to give its **image**. The object point is the same distance from the line as the image point.

EXERCISE **9.11**

Copy each diagram and draw the reflected image in the line.

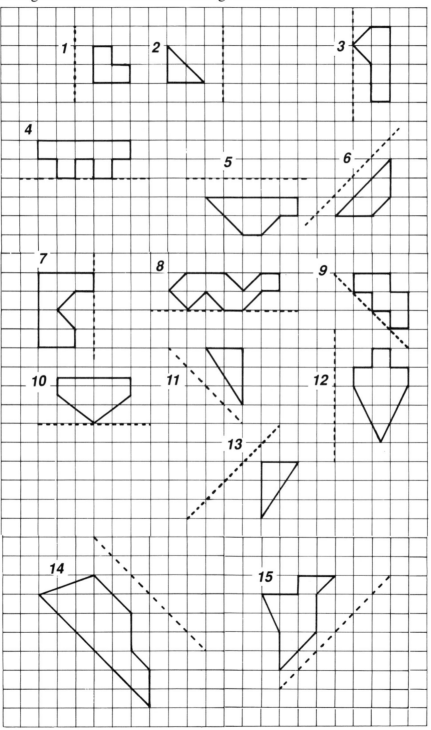

EXERCISE 9.12

For each question draw in the mirror line.

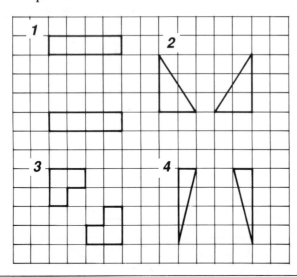
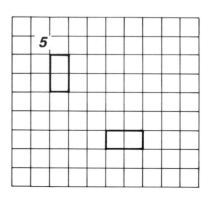
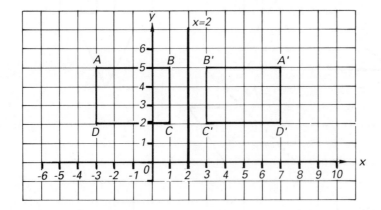

Example 3

ABCD has been reflected in the line $x = 2$ to give the image A'B'C'D'.

Write down the coordinates of points A,B,C and D, then the coordinates of A',B',C' and D'. Do you notice anything about the points when you compare how they have changed?

EXERCISE 9.13

1 Draw in the image of the shape as reflected (*a*) in the *x* axis, (*b*) in the *y* axis

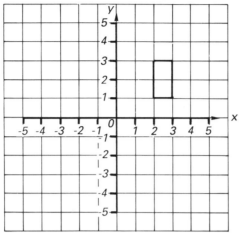

2 Draw in the image of the shape as reflected (*a*) in the *x* axis, (*b*) in the *y* axis.

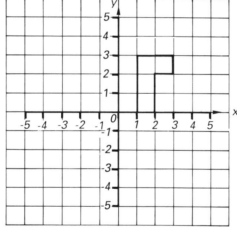

3 Draw in the image of the shape as reflected (*a*) in the *x* axis, (*b*) in the line *x* = 2.

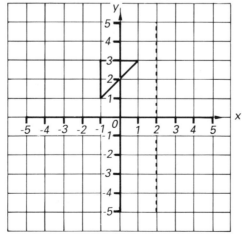

4 Draw in the image of the shape as reflected (*a*) in the *y* axis, (*b*) in the line $y = -1$.

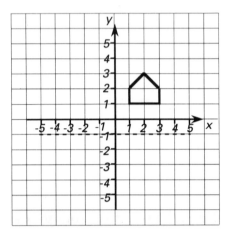

5 Draw in the image of the shape as reflected (*a*) in the *x* axis, (*b*) in the *y* axis, (*c*) in the line $y = x$.

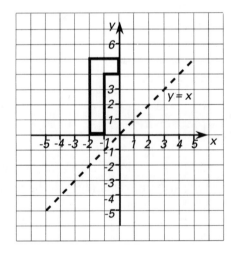

6 Draw axes labelled from -8 to 8. Plot the points (1,4), (1,7) and (4,7), joining them to make a triangle. Add to the diagram the lines $y = x$ and $y = -x$. Reflect the triangle in (*a*) the *x* axis, (*b*) the *y* axis, (*c*) the line $y = x$, (*d*) the line $y = -x$.

7 Draw axes labelled from -10 to 10. Plot the points $(-4,8)$, (1,8), (1,6), $(-2,6)$, $(-2,4)$ and $(-4,4)$. Reflect the shape in the lines $x = 3$ and $y = 2$.

Investigation K

In snooker a ball hits the cushion and rebounds, its path is a reflection of the first part of its path. Draw a scale diagram of a snooker table 6 ft × 3 ft, and use this idea of reflections to plot the path of balls as they rebound off the cushion.

Try a few examples with the ball starting in different places on the table.

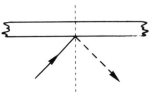
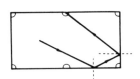

Rotational symmetry

In Book 1 you were introduced to rotational symmetry. A shape can fit precisely on top of itself, and look exactly the same as it rotates about a point. The number of times it does this while being rotated is called the **order** of rotational symmetry.

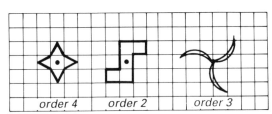

order 4   order 2   order 3

EXERCISE 9.14

For each shape, give the order of rotational symmetry.

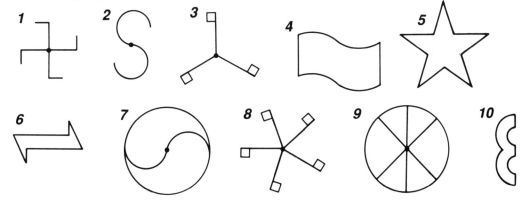

Use the shape given to construct a diagram which has a rotational symmetry of the order suggested.

order 4   order 5   order 3   order 2   order 6

Rotations

Shapes can be rotated about a point. Trace triangle ABC, and holding the tracing paper down on to the page at the point ×, rotate it one quarter ($\frac{1}{4}$) of a turn, or 90° *anticlockwise.* You will end up with the shape at position A'B'C'. This is the way shapes are rotated.

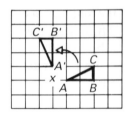

The rectangle has been rotated half ($\frac{1}{2}$) a turn or 180°. Check this with your tracing paper. Would we describe the rotation as 180° clockwise, or 180° anticlockwise, or does it matter?

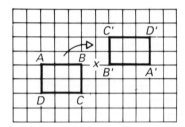

This is a rotation of $\frac{1}{4}$ turn or 90° clockwise about the origin. This rotation could also be described another way. Do you know what the rotation is?

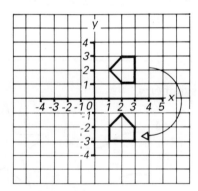

EXERCISE 9.15

Copy each shape and rotate it 90° clockwise about the point indicated.

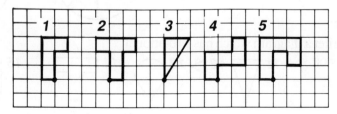

Find the angle of rotation, and describe the direction of rotation.

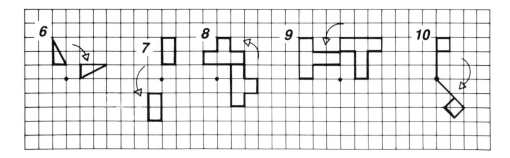

EXERCISE 9.16

Draw the object shape and add the image described.

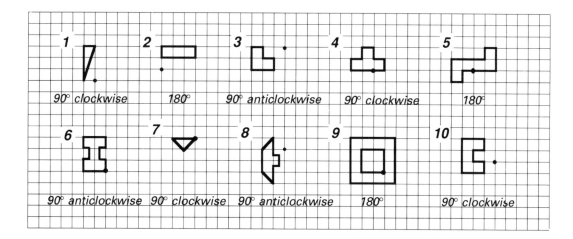

Investigation L

Find some shapes which have two lines of symmetry and rotational symmetry of order 2. Is there anything else these shapes have in common? Extend the problem to compare shapes with three lines of symmetry, and rotational symmetry of order 3, etc.

Revision exercises: Chapters 5-9

1 What is the number represented by the Roman numerals (*a*) XI (*b*) XIX (*c*) XCV (*d*) DCLI (*e*) LXII (*f*) CLX (*g*) MLII.

2 Write down the Roman numerals which would represent the numbers (*a*) 7 (*b*) 12 (*c*) 83 (*d*) 92 (*e*) 112 (*f*) 232 (*g*) 1080 (*h*) 2040 (*i*) 3221.

3 Which Roman numerals would represent the copyright date on the film *The Return of the Jedi* ©1983?

4 During which year was *University Challenge* ©MCMLX first shown on television?

5 Change these fractions into decimals: (*a*) $\frac{3}{8}$ (*b*) $\frac{3}{20}$ (*c*) $\frac{1}{16}$ (*d*) $\frac{3}{5}$ (*e*) $1\frac{3}{4}$ (*f*) $2\frac{4}{5}$ (*g*) $4\frac{3}{16}$ (*h*) $1\frac{5}{8}$.

6 Change these fractions into decimals: (*a*) $\frac{1}{5}$ (*b*) $\frac{7}{8}$ (*c*) $8\frac{1}{4}$ (*d*) $4\frac{9}{10}$ (*e*) $2\frac{1}{20}$ (*f*) $7\frac{29}{100}$.

7 Change these decimals into fractions: (*a*) 0.8 (*b*) 0.3 (*c*) 0.7 (*d*) 0.14 (*e*) 3.2 (*f*) 5.3 (*g*) 8.06 (*h*) 2.41 (*i*) 5.24.

8 Change these decimals into fractions: (*a*) 0.6 (*b*) 0.12 (*c*) 0.75 (*d*) 3.8 (*e*) 0.125 (*f*) 12.48 (*g*) 4.69 (*h*) 6.9 (*i*) 2.008.

9 Find the first five (*a*) square numbers, (*b*) triangle numbers.

10 How many square numbers are there between 50 and 100?

11 Find all the square numbers between 10 and 50.

12 Find the next three terms in each of the following number sequences:

    (*a*) 1, 2, 4, 12, 36, 144, …   (*b*) 5, 15, 26, 38, 51, …

    (*c*) 4, 2, 5, 1, 6, …   (*d*) 16 000, 4000, 1000, …

    (*e*) 10, 8, 5, 1, …   (*f*) −1, 2, −4, 8, −16, …

    (*g*) 32, 320, 2880, 23 040, …   (*h*) 10, 20, 11, 19, …

13 In each case write down the new temperature:

(*a*) The temperature falls 4°C from 3°C during the night.

(*b*) A temperature of −4°C rises by 3 degrees.

(*c*) A temperature of 1°C falls by 2 degrees.

(*d*) A thermometer showing 8°C rises 6°C when moved into the sunlight.

(*e*) The temperature rises 4 degrees from −1°C.

14 On a shelf there are two books: one is blue and the other is red. In how many different ways can these two books be arranged? How many ways can three books be arranged once a green book is added? Continue the investigation for 4 and then 5 books, presenting your answers in a table. Find out how many ways you could arrange (*a*) 8, (*b*) 10, (*c*) 12 books.

15

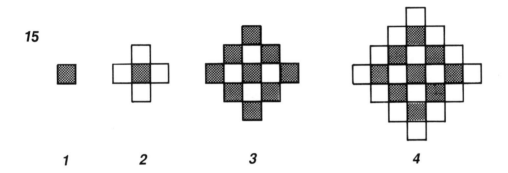

1   2   3   4

Diagram	1	2	3	4	5	6
White squares	1	4	4			
Shaded squares	0	1	9			
Total squares	1	5	13			

Complete the table for the 4th, 5th and 6th diagrams. By spotting a pattern in the numbers, find out how many (*a*) white, (*b*) coloured squares will be needed for diagrams (i) 10, (ii) 12, (iii) 15.

16 Children's blocks are stacked up against a wall. Count up how many blocks there are for 1, 2 and 3 levels as shown in the diagram. How many will there be for 4 and 5 levels? How many will there be for (*a*) 8, (*b*) 12, (*c*) 15 levels?

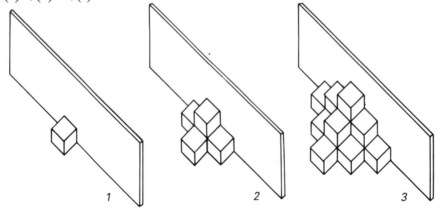

17 Diagram 1 has 2 roots, diagram 2 has 4 roots, and diagram 3 has 8 roots. How many roots will there be for diagrams (*a*) 5, (*b*) 10, (*c*) 15?

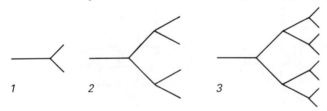

18 Cardboard rolls 10 cm in diameter are stacked as shown. How many could be stacked in a space 40 cm wide? Find out how many could be stacked in a space of width (*a*) 80 cm, (*b*) 120 cm, (*c*) 150 cm.

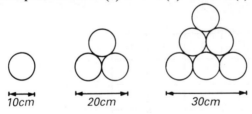

19 The diagram shows the stages in building up a pattern while laying out a tiled floor, based on a design found in an Arab building.

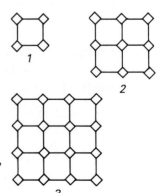

Stage	1	2	3	
Octagonal tiles	1	4		
Square tiles	4	9		

Complete and extend the table. Work out how many (*a*) octagonal, (*b*) square tiles would be needed for stages (i) 7, (ii) 10, (iii) 15.

20 The table gives the fuel consumption of a car, in kilometres per litre, at various speeds. Draw a bar graph of this information.

Speed (km/h)	30	50	80	100	110
Fuel consumption (km/litre)	16	19	14	12	11

21 A survey of colours of cars has the following information.

Colour	white	yellow	green	red	blue	black	other
Number of cars	78	35	47	83	67	59	47

Draw a bar graph to show this information.

22 The diagram gives the approximate distances in miles of various places from London.

Distance from London (miles)

(*a*) Which town is almost 300 miles from London?

(*b*) Estimate how far Norwich is from London.

(*c*) Which two places are the same distance from London?

(*d*) Which town is twice as far away as Lincoln is from London?

(*e*) Which town is about half the distance Shrewsbury is from London?

(*f*) Which town is the nearest to London?

23 Attendances at five test-match cricket grounds are shown in the diagram.

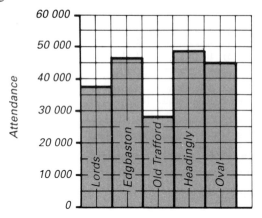

(*a*) Estimate the number of spectators at Headingly.

(*b*) What is the total attendance at all five matches?

(*c*) One of the five days at Old Trafford was washed out. Estimate what the attendance would have been on that day assuming that on each of the other four days the number of spectators was the same.

24 A class of 25 pupils conducted a survey to investigate the shoe sizes worn by the members of the class. The results are shown in the diagram.

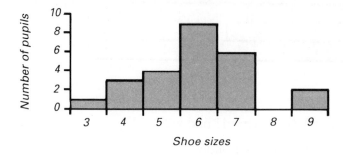

(*a*) How many wore size 4?

(*b*) How many wore size 8?

(*c*) How many had a shoe size of under 6?

(*d*) How many had a shoe size of over 6?

25 Five players in a netball team have scored this season, as shown in the bar chart.

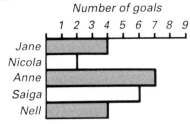

Number of goals

(*a*) How many goals have they scored altogether?

(*b*) Which two have scored the same number of goals?

(*c*) In their next game Nell scores 3 goals, Saiga, Anne and Nicola score one each. Who is the top scorer now?

26 (*a*) An insurance company plan guarantees a certain sum, depending on monthly payments as shown in the table.

Monthly payment	£10	£20	£30	£50	£75
Guaranteed sum	£1300	£2500	£4000	£7000	£10 000

Draw a bar chart to show this information.

(*b*) On top of the guaranteed sum, the company predicts the extra bonus likely to be earned as shown in the table.

Monthly payment	£10	£20	£30	£50	£75
Extra bonus	£2000	£4500	£7000	£11 500	£17 000

By using a different shading, or colour, extend your bar graph to show the bonus to be added to the guaranteed sum.

27 Five girls are given pocket money each week: Jane has £2, Nicola £1.80, Anne 90p, Saiga and Nell both have £1.20. Draw a bar chart to represent this information.

28 The pie chart shows the number of pupils who preferred each colour of exercise book. If 9 preferred blue, estimate how many preferred (*a*) red, (*b*) yellow, (*c*) green.

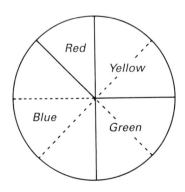

29 The shaded area of the pie chart represents £20. How much does each of the other sectors represent?

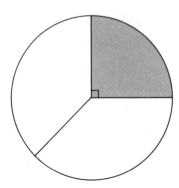

30 In the pie chart, sectors A and B are equal, sector C is a quarter of the diagram, and sector D is a third. If sector C represents 60 people, how many people are represented by (*a*) sector A, (*b*) sector B, (*c*) sector D?

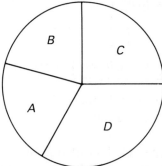

31 For each £100 earned by a small company they pay out the money as shown by the pie chart.

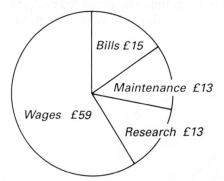

(*a*) If the company earned £1500 in one week, how much could be passed on in wages?

(*b*) How much would the company need to earn in order to have £650 available for research?

32 This is a portion of a train timetable for services on a Saturday.

Saturdays

LIVERPOOL CENTRAL	1211†			1311†	1411†			1511†	1611†			1711†				1826†	
KIRKBY	1229			1329	1429			1529	1629			1729				1849	
SOUTHPORT		1215	1324			1415	1524			1615	1715		1820	1830			
WIGAN WALLGATE arr.	1255	1257	1352	1355	1455	1457	1552	1555	1655	1657	1755	1757		1845		1912	1915
WIGAN WALLGATE dep.	1258	1302	1358	1402	1458	1502	1558	1602	1658	1702	1758	1802	1827	1847	1847	1917	1920
Ince	1302	1305	1402	1405	1502	1505	1602	1605	1702	1705	1802	1805	1830	1850	1850	1920	1923
Hindley	1306	1309	1406		1506	1509	1606	1609	1706	1709	1806	1809	1834	1854	1854	1924	1927
Westhoughton	1311		1411		1511		1611		1711		1811		1840	1900	1900		1933
BOLTON	1320		1420		1520		1620		1724		1820		1848	1910	1910		1941
Daisy Hill		1314		1414		1514		1614		1714		1814				1929	
Hag Fold		1317		1417		1517		1617		1717		1817				1932	
Atherton		1319		1419		1519		1619		1719		1819				1934	
Walkden		1325		1425		1525		1625		1725		1825				1940	
Moorside		1329		1429		1529		1629		1729		1829				1944	
Swinton		1332		1432		1532		1632		1732		1832				1947	
Pendleton		1337		1437		1537		1637		1737		1837				1952	
Salford Crescent		1340		1440		1540		1640		1740		1840				1955	
Salford		1344		1444		1544											
MANCHESTER VICTORIA	1337	1347	1437	1449	1537	1547	1637	1647	1741	1747	1837	1847	*1916*	1927	1927	2001	*2027*

LIVERPOOL CENTRAL					
KIRKBY					
SOUTHPORT		1945	2045	2145	2250
WIGAN WALLGATE arr.		2027	2117	2227	2328
WIGAN WALLGATE dep.	2007	2028	.2117	2228	2328
Ince	2010	2032	2120	2232	2332
Hindley	2014	2036	2124	2236	2336
Westhoughton	2020	2041	2130	2241	2341
BOLTON	2027	2050	2140	2250	2350
Daisy Hill					
Hag Fold					
Atherton					
Walkden					
Moorside					
Swinton					
Pendleton					
Salford Crescent	*2059*		2200		
Salford					
MANCHESTER VICTORIA	*2057*	2111	2207	2310	0007

(*a*) How many trains call at Swinton during this part of the day?

(*b*) How many trains which leave Wigan Wallgate go to Bolton?

(*c*) Which train would you catch from Ince to be in Atherton by 1530?

(*d*) What is the last direct train you could catch from Kirkby to Pendleton?

(*e*) How many trains are there in Wigan Wallgate at 1556?

(*f*) Why do you think this is?

(*g*) Amanda catches the 1415 Southport train and wants to get to Bolton as quickly as possible. Describe how she would do this, and the time she would arrive in Bolton.

(*h*) Omar arrives at Kirkby station at 1630. What is the earliest time he could be in Manchester?

(*i*) Marion travels regularly between Kirkby and Walkden in the afternoons, sometimes having to change trains on the way. Make out a timetable for Marion which shows arrival times at Walkden for each train leaving Kirkby.

33 The timetable shown is for buses between Bury and Trafford Park.

BURY—TRAFFORD PARK via Unsworth, Prestwich, Lower Kersal, Pendleton and Ordsall   **92**
Includes Service 732

Mondays to Fridays

BURY, Interchange	0625		0715		0745		0815	0845		1445		1515	1550	1600	1615	1630	1645	1700
Unsworth, Pole	0639		0731		0801		0831	0901		1501		1531	1606	1616	1631	1646	1701	1716
Besses o'th' Barn Station	0647		0741		0811		0841	0911		1511		1541	1616		1641		1711	
Prestwich, Tower Buildings	0649		0745		0815		0845	0915	AND	1515		1545	1621		1645		1715	
Carr Clough, Butterstile Lane/Sandy Lane	0654		0751		0821		0851	0921	EVERY	1521	╲	1551	1626		1651		1721	
Agecroft, Kersal Vale	0659	0728	0758	0818	0828	0848	0858	0928	30	1528	1545	1558	1633		1658		1728	
Lower Kersal, Flats	0702	0731	0801	0821	0831	0851	0901	0931	MINS.	1531	1548	1601	1636		1701		1731	
Pendleton Precinct arr.	0711	0742	0812	0832	0842	0902	0912	0942	UNTIL	1542	1559	1612	1647		1712		1742	
Pendleton Precinct dep.	0711	0744	0814		0844		0914	0944		1544		1614	1647		1714		1744	
Trafford Road/Regent Road	0716	0750	0820		0850		0920	0950		1550		1620	1653		1720		1750	
ORDSALL, Salford Quays		0755	0825		0855		0925	0955		1555		1625	1658		1725		1755	
Trafford Bridge	0721																	
TRAFFORD PARK, Third Avenue	0727																	

BURY, Interchange	1715	1745		1815		1915			
Unsworth, Pole	1731	1801		1831		1931			
Besses o'th' Barn Station	1741	1811		1841		1941			
Prestwich, Tower Buildings	1745	1815		1845		1945			
Carr Clough, Butterstile Lane/Sandy Lane	1751	1821		1851		1951			
Agecroft, Kersal Vale	1758	1828	1833	1858	1933	1958	2033	2133	2233
Lower Kersal, Flats	1801	1831	1836	1901	1936	2001	2036	2136	2236
Pendleton Precinct arr.	1812	1842		1912		2012			
Pendleton Precinct dep.	1814		1845		1945		2045	2145	2245
Trafford Road/Regent Road	1820		1850		1950		2050	2150	2250
ORDSALL, Salford Quays	1825		1855		1955		2055	2155	2255

⊕—Schooldays only, shows Service Number 732.   △—Commences from Cavendish Road School 5 minutes earlier.

(*a*) Which bus would you catch from Butterstile Lane to be at work on Regent Road by 0830?

(*b*) Mark lives at Prestwich and wants to go shopping in Pendleton Precinct no later than 1630. What is the latest time he needs to be at his bus stop?

(*c*) Dave and Sue live in Agecroft and want to see a film starting at 7.30 p.m. at the Salford Quays Cinema Complex. At what time should they be at their bus stop?

(*d*) Throughout the whole day, how many buses stop at Unsworth?

(*e*) Gill arrives by train at Besses station at 1730, and wants to go to Salford Quay by bus. Describe her journey, and the earliest time she can be at her destination.

(*f*) How long is the journey from Bury on the 1645 to Pendleton Precinct?

34 A drawing of a garden has a scale of 1 to 100. Complete the table.

	Length	Width	Fencing	Wall	Patio	Pond	Rose bed
Distance on drawing		22 cm	47 cm		1.8 cm		
Actual distance	27 m			18.4 m		2.4 m	3.7 m

35 A drawing 330 mm long represents a boat to a scale of 1 : 30. (*a*) How long is the boat, in metres? (*b*) If the real width of the boat is 3.6 metres, how wide is the drawing?

36 The distance from Appleby to Whitam is 6 cm on a map with a scale 1 : 250 000. How far apart are the two towns in kilometres?

37 A toy car is 36 mm long. The real car is 3.6 m long. To what scale has the toy car been made?

38 You are drawing a bar chart to represent the heights of mountains, using a scale of 1 to 500 000. Complete the table, giving your answer to the nearest millimetre.

Mountain	Kilimanjaro	Kenya	Mont Blanc	McKinley	Everest
Height (m)	5900	5200	4800	6200	8850
Length of bar (mm)					

39 (*a*) In a drawing of a park with a scale of 1:150, how long would a path be, in millimetres, which was actually 40 m long?

(*b*) The football pitch in the park is 100 m long. How long will this have to be on the drawing?

40 A company has a sliding scale of prices depending on how many of their wooden battens are bought.

Battens	10	20	30	40	50	60
Cost	£6.00	£10.50	£15.00	£19.50	£24.00	£28.50

Draw a graph of this information, and use your graph to answer the following questions.

(*a*) How much would it cost you to buy (i) 36 battens, (ii) 52 battens.

(*b*) Find the number of battens you would get for (i) £7.00, (ii) £13.00.

41 Draw a conversion graph to help change a maximum of £3 into drachmae, if the conversion rate is £1 = 250 drachmae.

(*a*) Change these amounts of money into drachmae: (i) £2.50 (ii) £1.60.

(*b*) Change these drachmae into sterling: (i) 550 d (ii) 75 d.

42 Draw a conversion graph to help change a maximum of £5 into D-marks, if the conversion rate is £1 = 3.10 D-marks.

(*a*) Change these amounts of money into D-marks: (i) £2.50 (ii) £4.00.

(*b*) Change these D-marks into sterling: (i) 6.20 D-marks (ii) 7.75 D-marks.

43 Draw a conversion graph to help change a maximum of £5 into I-punts, if the conversion rate is £1 = £1.20 I-punts.

(*a*) Change these amounts of money into I-punts: (i) £2.50 (ii) £4.25.

(*b*) Change these I-punts into sterling: (i) £4.20 I-punts (ii) £1.50 I-punts.

44 A copper tube 1 metre long costs £1.50 to produce. Draw a graph to help calculate how much it would cost to produce tubes of up to 5 metres in length. Use your graph to answer these questions:

(*a*) How much would it cost to produce a copper tube of length (i) 3.5 m, (ii) 4.8 m?

(*b*) What length of tube could be produced for (i) £6.30, (ii) £3.90?

45 Electric cable costs 45p per metre. Draw a graph to show how much it would cost for any length up to 100 metres.

(*a*) How much would it cost for a length of (i) 55 m, (ii) 85 m?

(*b*) What length could you buy for (i) £5.00, (ii) £10.50?

46 The graph represents the journey made by two people out for a drive in their car. While out they stopped at a café for lunch, and had to stop for a short time when the car broke down.

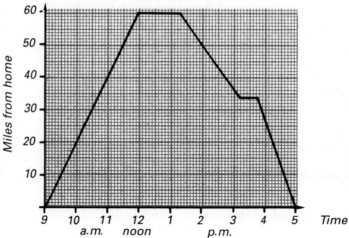

(*a*) When did they stop for lunch?

(*b*) How far from home were they?

(*c*) How long did they stay at the café?

(*d*) When did the car break down?

(*e*) How many miles was the entire journey?

47 Jennifer went hiking for a day, stopping each time she passed a boating lake. The graph represents her journey.

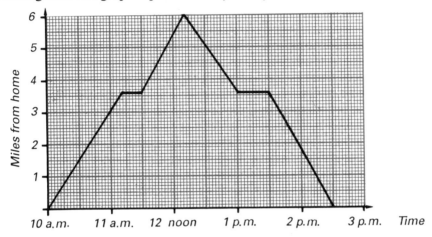

(*a*) How far was the boating lake from her home?

(*b*) What happened when she was 6 miles from home?

(*c*) What time was it when she first stopped?

(*d*) What was the total time she spent at the boating lake?

(*e*) For how long was Jennifer actually walking?

48 The graph represents a record of the journey of a motorcyclist.

(*a*) How far was she from home when she stopped?

(*b*) For how long did she stop?

(*c*) At what time did she arrive home?

(*d*) How far did she travel on the bike?

(*e*) What part of the journey was the fastest?

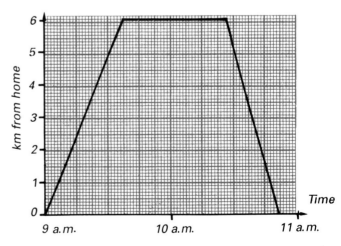

49 The graph shows the temperature as it rose during a particular day.

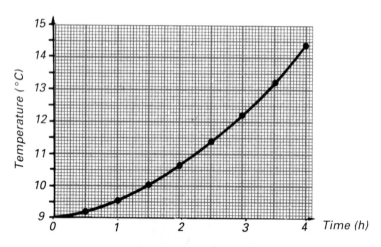

(*a*) How often were the temperature readings taken?

(*b*) What was the maximum temperature reached after the 4 hours?

(*c*) How far did the temperature rise during the (i) second hour, (ii) third hour?

(*d*) What was the temperature after (i) $1\frac{3}{4}$ hours, (ii) $3\frac{1}{4}$ hours?

(*e*) After how long was the temperature (i) 11°C, (ii) 13.6°C?

50 (*a*) What was the highest share price?

(*b*) During which month was the price most stable?

(*c*) What was the price on 1st February?

(*d*) What was the lowest price of the shares?

(*e*) During which month did the price rise the most?

(*f*) During which month did the price fall the most?

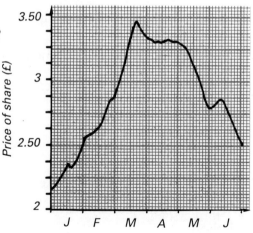

51 These sketch graphs show the distance of Kevin from his dog Patch as they play in the park. Match one of the graphs with each of these descriptions.

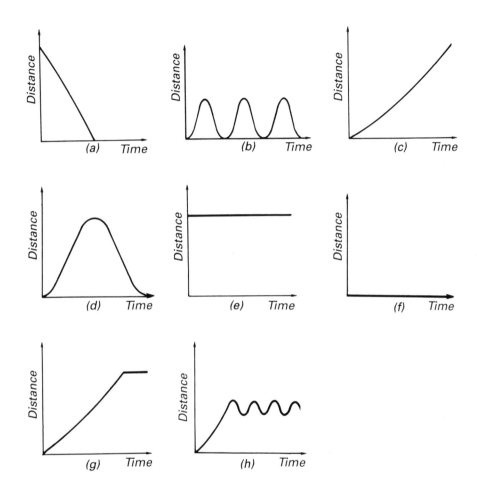

(i)   Patch runs off to fetch a stick and bring it back.

(ii)   Patch races across the field and stops on the other side.

(iii)   Patch refuses to move from where he has stopped.

(iv)   Patch returns to Kevin and sits down in front of him.

(v)   Patch runs to fetch the stick, but jumps to and fro over it.

(vi)   Patch repeatedly runs a short distance away from Kevin and returns to him.

(vii)   Patch runs off across the field.

(viii) Patch refuses to move from Kevin's side.

52 Draw the following shapes accurately.

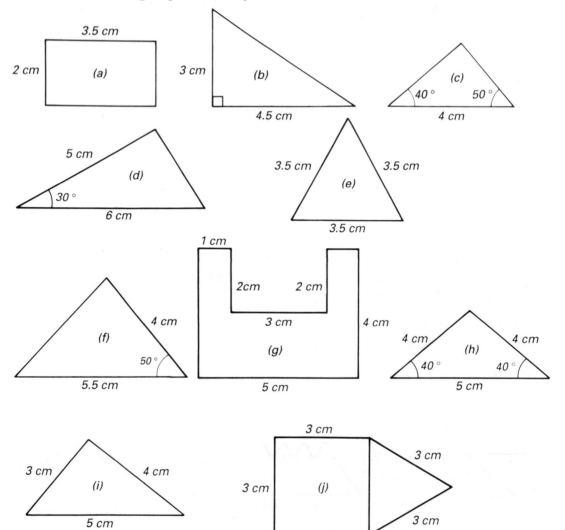

53 In each question you are given two of the angles of a triangle; work out the third angle in each case:

(a) 82°, 20°   (b) 60°, 60°   (c) 70°, 40°   (d) 70°, 70°   (e) 36°, 58°

(f) 27°, 53°   (g) 41°, 49°   (h) 86°, 82°   (i) 178°, 1°   (j) 123°, 45°

(k) 30°, 30°   (l) 68°, 86°   (m) 23°, 73°   (n) 52°, 52°   (o) 37°, 51°

54 A square is cut in half. What are the three angles of one of the triangles formed?

55 In each of the following, work out angle *x*.

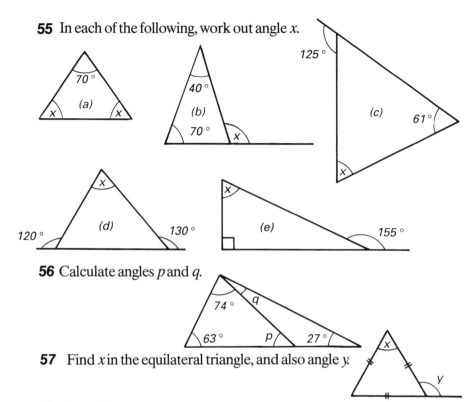

(a) 70°

(b) 40° 70° *x*

(c) 125° 61° *x*

(d) *x* 120° 130°

(e) *x* 155°

56 Calculate angles *p* and *q*.

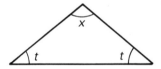

74° *q* 63° *p* 27°

57 Find *x* in the equilateral triangle, and also angle *y*.

x *y*

58 A roof slopes so that the angle at *x* is 110°. What is angle *t*?

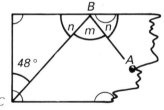

x *t* *t*

59 A snooker ball hit from point A rebounds off the cushion at B to end up in pocket C. Calculate angles *m* and *n*.

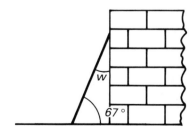

B *n* *m* *n* 48° A C

60 A ladder leans against a wall as shown. What is angle *w*?

w 67°

61 In each of the following questions you are given three angles of a quadrilateral; work out the fourth angle in each case.

(*a*) 90°, 90°, 90°   (*b*) 120°, 130°, 40°   (*c*) 74°, 39°, 170°

(*d*) 100°, 79°, 89°   (*e*) 130°, 130°, 50°   (*f*) 83°, 131°, 69°

(*g*) 150°, 150°, 59°   (*h*) 60°, 80°, 100°   (*i*) 160°, 60°, 90°

62 ABCD is a square which has triangle ACD replaced by equilateral triangle ACE. Calculate the size of angle BAE.

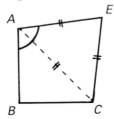

63 In each of the following work out angle *x*.

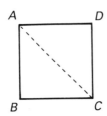

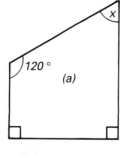

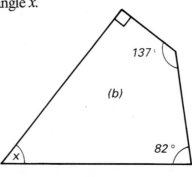

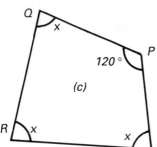

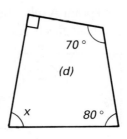

64 Work out angles *x*, *y* and *z* in this trapezium with two parallel sides.

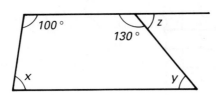

65 This house end has three right angles and two other angles, both equal. Find the other two angles.

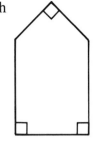

66 If angle x is 75°, work out angle y.

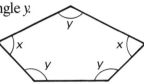

67 Angle a is 270°, angle b is half of angle c. Work out angles b and c.

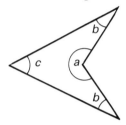

68 Draw the following diagrams to the scale stated, writing the new measurements on them.

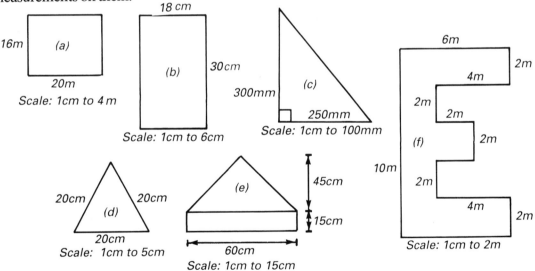

(a) 16m, 20m
Scale: 1cm to 4 m

(b) 18 cm, 30cm
Scale: 1cm to 6cm

(c) 300mm, 250mm
Scale: 1cm to 100mm

(d) 20cm, 20cm, 20cm
Scale: 1cm to 5cm

(e) 45cm, 15cm, 60cm
Scale: 1cm to 15cm

(f) 6m, 2m, 4m, 2m, 2m, 2m, 10m, 2m, 4m, 2m
Scale: 1cm to 2m

69 The diagram shows a floor plan of an office. Find

(*a*) the width of the office,

(*b*) the length of the office,

(*c*) the distance across the office from corner to corner,

(*d*) the shortest distance from the telephone socket to a wall.

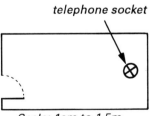

telephone socket

Scale: 1cm to 1.5m

70 For each room in the house find (*a*) the length, (*b*) the width, (*c*) the area.

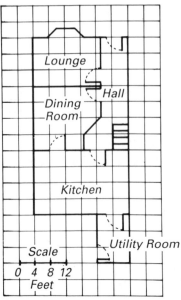

71 Draw an enlargement of each diagram to the given scale.

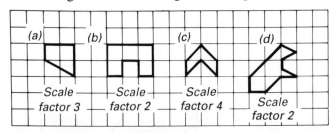

72 Copy each diagram, and draw the enlargement to the given scale factor from the point of enlargement.

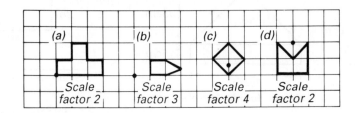

73 How many lines of symmetry have each of the following shapes?

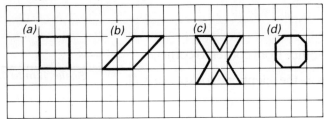

74 Copy each shape and show the reflected image in the line.

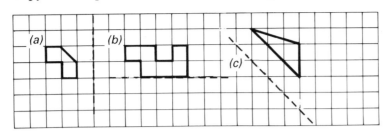

75 Plot each set of coordinates on a pair of axes, complete each shape and draw the enlargement to the given scale factor, using (0,0) as the point of enlargement.

(*a*) A(1,1), B(1,3), C(−3,3), D(−3,1), scale factor 2.

(*b*) A(0,0), B(−1,2), C(−2,0), D(−1,−2), scale factor 4.

(*c*) A(2,1), B(−1,3), C(0,1), D(−1,−1), scale factor 3.

76 Copy and draw in the image as reflected (*a*) in the *x* axis, (*b*) in the line *x* = −1, (*c*) in the line *y* = *x*.

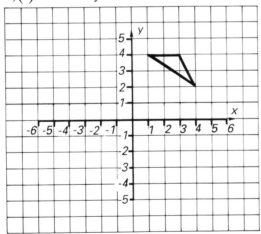

77 Give the order of rotational symmetry for each of the shapes in question 73.

78 Find the angle of rotation and describe the direction of rotation for each of these diagrams.

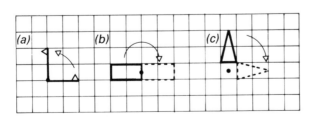

79 Copy each object shape and add the image described.

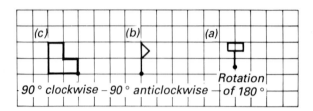

10. Measuring: decimals

Distances

Example 1

A ⊢————————————————⊣ B

Measure the length of the line AB in millimetres.

It is 68 millimetres long. We usually write 68 mm.

As there are 10 millimetres in one centimetre, we can divide 68 mm by 10 to give us the length of the line in centimetres; 6.8 cm. (You should not need your calculator to divide by 10! – if you have forgotten how to do this, then check back to Chapter 1 on Decimals: division).

 We usually measure short distances in **centimetres** or **millimetres**, e.g. things like the size of a page in a book, dimensions of a brief-case, etc. For longer lengths, we usually measure them in **metres**. Remember that

 100 centimetres = 1 metre
 10 millimetres = 1 centimetre
 1000 millimetres = 1 metre

(As a rough guide, a classroom doorway is about 2 m high. Also, you may well have used a metre stick in your classroom.)

Example 2

Measure the length and width of this rectangle, in centimetres. How far is it along all four sides (i.e. the *perimeter*)?

Length = 6.3 cm
 Width = 3.2 cm
Distance all round (the perimeter)
 = 6.3 cm + 3.2 cm + 6.3 cm + 3.2 cm
 = 19.0 cm

Example 3

In this triangle, how much shorter is it along the diagonal from A to C than it is from A to B then on to C?

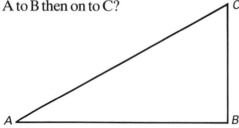

AB = 59 mm and BC = 32 mm.

So   A to B then on to C = 59 mm + 32 mm
$$= 91 \text{ mm}$$

AC = 66 mm.
So AC is shorter by

$$91 \text{ mm} - 66 \text{ mm} = 25 \text{ mm}$$

EXERCISE 10.1

1 How far is it around this triangle?
   Work (*a*) in millimetres, (*b*) in centimetres.

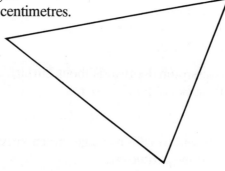

2 Measure the width and the height of the page you are now
   writing on. Give each answer in (*a*) millimetres, (*b*) centimetres.

3 Draw a rectangle 8 cm long and 6 cm high, as accurately as you
   can. Draw a diagonal, and then measure its length.

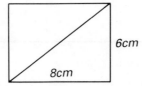

6cm

8cm

(We can show that it should be exactly 10 cm – how near were you?)

4 In this model car, measure (*a*) the length of the car, (*b*) its height, (*c*) the wheel diameter, giving your answers in millimetres.

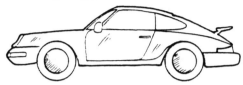

5 Draw a triangle which has one side 56 mm long and another side 83 mm long, with an angle of 73° between these sides. (Check back to Chapter 8 if you can't remember how to draw an angle.)

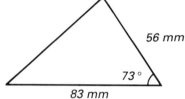

56 mm

73 °

83 mm

6 Make an accurate drawing of this diagram of the end of a rack which holds cassettes.

Measure the sloping edge AB.

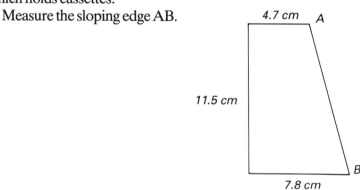

4.7 cm   A

11.5 cm

B

7.8 cm

7 A computer disc box measures 165 mm by 155 mm, and is 38 mm deep. Make an accurate drawing of (*a*) the top, (*b*) the longer side.

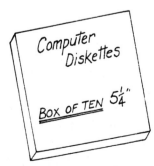

Computer Diskettes

BOX OF TEN 5¼″

EXERCISE 10.2

Now try some further questions on distances.

1 I make two car journeys, one of 23.7 km and another of 18.9 km. How far is the total distance travelled on these two journeys?

2 The mileage counter on my car read 33 512.7 this morning. Tonight it read 33 594.3 when I arrived home. How many miles had I travelled during the day?

3 PQ, QR and PR are paths across a park. How much shorter is it to walk across the path PR, than to walk along PQ then QR?

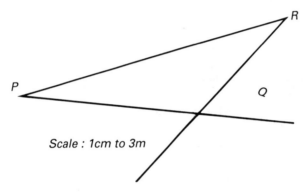

Scale : 1cm to 3m

4 I need to buy skirting board for a new bedroom I have built. The room is rectangular, and measures 3.82 m by 2.94 m. The bedroom has two doors, each 0.76 m wide. What length of skirting board do I need?

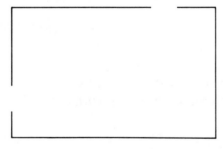

5 One one day a sales representative drove from Southampton to Bristol, then to Birmingham and on to London before returning to Southampton. According to her road atlas the four journeys are 76 miles, 101 miles, 105 miles and 77 miles. When she left home in the morning the mileage counter read 3556 miles, and when she returned in the evening the reading was 3897 miles. Work out (*a*) the actual distance she drove, (*b*) the distance according to the road atlas, (*c*) the difference in distances (*a*) and (*b*).

Yards, feet and inches

Your parents and grandparents will often refer to distances in yards, feet and inches, as they were the units used 'in the good old days' when they were at school!

Before going 'metric', distances were measured in 'imperial' units. Small distances were measured in **inches** – one inch is about 25 millimetres ($2\frac{1}{2}$ cm). There are 12 inches in a **foot** which is about 300 mm (or 30 cm) and 3 feet in a **yard**.

So a yard is a little shorter than a metre.

Most reference tables, and some calculators, compare metric and imperial units as follows

> 1 inch = 25.4 mm
> 1 foot = 12 inches = 305 mm (30.5 cm)
> 1 yard = 3 feet = 914 mm (91.4 cm)
> 1 metre = 39.37 inches

Longer distances are measured in **miles**; 1 mile = 1760 yards, which is about 1.6 km. (As a rough estimate, 8 km = 5 miles.)

Investigation A

Try to find out how the 'old' units of distance came about. (It is likely, for example, that a foot was the length of a man's foot.) There are other distances as well – a chain, a fathom, a hand, etc.

Extension

How was a metre defined, orginally? What is the definition today?

EXERCISE 10.3

1 In building a room divider, wooden beams 4 inches wide and 2 inches thick are used. Draw a cross-section of a beam. (You need a ruler marked in inches.)

2 A basketball player is 6 feet 9 inches tall. How many inches is this?

3 The world record for spitting a melon seed was 37 feet 4 inches in 1971.

  (*a*) Write this distance in yards, feet and inches.

  (*b*) How many inches is this altogether?

4 A plank of wood 5 feet long is cut into six equal pieces to make shelves. How long is each shelf, in inches? Draw a plan of a shelf, given that it is $4\frac{1}{2}$ inches wide.

5 The tallest giraffe was 4 feet 6 inches taller than a double decker bus, which was 14 feet 6 inches high. How tall was the giraffe?

6 I need a fence post 5 feet 7 inches long, and I have one which is 8 feet 2 inches long. How much will be left over when I cut it?

7 The 'Titanic' liner was 882 feet long. How many yards is this?

8 Mount Everest is estimated to be 5.5 miles above sea level. How many feet is this, correct to the nearest thousand? (3 feet = 1 yard and 1760 yards = 1 mile.)

9 When I measured the windows in my living room, I found that I needed 90 inches of curtain material. How many yards is this?

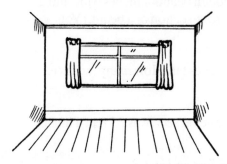

10 Measure the page you are writing on, giving its height and width in inches, as accurately as you can.

Weights

We weigh small, light objects in **grams**. Heavier objects are weighed in **kilograms**. Remember that 1 kilogram = 1000 grams. We can write this, in short, as

$$1 \text{ kg} = 1000 \text{ g}$$

Sugar usually comes in bags weighing one kilogram; a jar of jam weighs about half a kilogram, which is the same as 500 g. An average man weighs about 75 kg.

Example 4

A recipe for a cake requires 500 g of flour, 125 g of lard, 125 g of margarine, 150 g of sugar and 20 g of mixed peel.

(*a*) How heavy are the ingredients altogether?

(*b*) How heavy, in kilograms, would the ingredients for three cakes be?

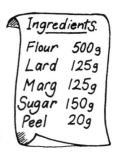

(*a*) Total weight = 500 g + 125 g + 125 g + 150 g + 20 g
      = 920 g

(*b*) For three cakes, the weight will be 3 × 920 g
                = 2760 g
                = 2.76 kg   (dividing *grams* by 1000 will give *kilograms*)

EXERCISE 10.4

1 There are 630 g of flour in a bag. If my sister calls with a 1.5 kg bag of flour, how much do we have between us?

2 If I take 860 g of sugar from a bag containing 1.6 kg, how much sugar is left in the bag?

3 In a 'litter-pick', three brothers collect 2.54 kg, 1.32 kg and 960 g of litter. How much have they collected altogether?

4 Ingredients for 24 buns weigh 1296 g. How much will each bun weigh, on average?

5 A tin of fruit weighs 470 g. How heavy will 12 tins be? Give your answer in kilograms.

6 If a packet of 80 tea-bags weighs 240 g, how much will one tea-bag weigh?

7 A can of beans weighs 450 g. (*a*) How much will a pack of four tins weigh? (*b*) How heavy will a box of 24 tins be, in kilograms?

8 In a maternity hospital a mother has triplets, weighing 2.65 kg, 2.07 kg and 1.34 kg. What is the total weight of the babies?

9 A wooden block weighs 286 g. How many of these blocks would I need if I have to have a total weight of at least 2.2 kg?

10 A large cheese weighs 5.63 kg. How many 400 g portions can be cut from this cheese?

Capacities

Although milk is normally sold in pints, most liquids (lemonade, washing-up liquid, etc.) are bought in **litres** or **millilitres** (ml). There are 1000 millilitres in one litre.

Example 5

A recipe requires 260 ml of pure orange juice. The local shop sells the orange juice in cartons containing 75 ml. How many cartons will you need, and how much juice will be left over?

You will need to find how many cartons (i.e. 75 ml) there are in 260 ml.

     i.e. $260 \div 75 = 3.466\,666\ldots$

This means that you will need more than 3 cartons, so you will have to buy 4 cartons.
   Now 4 cartons will contain $4 \times 75\,\text{ml} = 300\,\text{ml}$.
So there will be $300\,\text{ml} - 260\,\text{ml} = 40\,\text{ml}$ of juice left over.

Example 6

A wine bottle holds 750 ml. How many bottles will you need if you have 5 litres of home-made wine to bottle?

WINE

You will need to see how many bottles (i.e. 750 ml) there are in 5 litres.
   If you work out $5 \div 750$ on your calculator, you should get
$0.006\,666\,66\ldots$, so there is something wrong somewhere.
   Before you begin to do the division, you must have *both* capacities in the same units – either *both* in millilitres or *both* in litres. Try each way, to check that the answer is the same whichever way you do it.

In millilitres
$5000 \div 750 = 6.6666\ldots$

In litres
$5 \div 0.750 = 6.6666\ldots$

So you will need *seven* bottles (the seventh will not be full).

EXERCISE **10.5**

1 A bottle contains 1.5 litres of lemonade. How many glasses, each holding 250 ml can be filled from the bottle?

2 A bottle of medicine holds 80 ml. How many doses, each of 5 ml, should you be able to pour from the bottle?

3 'Buy our small bottle of washing-up liquid for 50p, or save by buying the large one for £1.50.'
    If the small bottle contains 750 ml, and the large one contains 2.5 litres, do you actually save by buying the large one?
    If so, how much extra liquid do you get?
    If not, how much less do you get?

4 A carton of milk contains 1.136 litres. How many full glasses of milk, each holding 200 ml, can be poured from the carton? How much milk is left?

5 A litre is about $1\frac{3}{4}$ pints. How many litres of milk do I buy in a week, if I buy one pint each day for seven days?

6 Using the information in question 5, how many pints will there be in a 25-litre barrel of beer?

7 A can of lemonade contains 330 ml. I buy six cans. My friend buys a 2 litre bottle of lemonade, saying that there is more in her bottle than in my six cans. Work out whether or not she is correct.

8 When I change the oil in my car, it takes three cans, each containing 440 ml of oil, to fill it up again. How many oil changes could I make from a 5-litre drum of oil?

9 It takes 20 drops of medicine to make a 5 ml dose. How much medicine is there in one drop?

10 A chemist has a box of twelve bottles, each containing 130 ml of a chemical. How many bottles can he take out of the box, if he has to keep at least half a litre of the chemical in the box?

Adding simple fractions

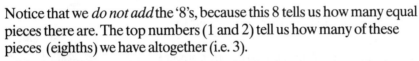

At a party a round chocolate cake has been cut into eight equal pieces. Ann has a piece, and Mike has two pieces. What fraction of the cake has been eaten?

Each piece is one eighth ($\frac{1}{8}$) of the whole cake. Ann and Mike have eaten three of these eighths, or $\frac{3}{8}$.

The fraction of the cake which is left is $\frac{5}{8}$.

We can write the fractions for Ann and Mike like this

$$\frac{1}{8} + \frac{2}{8} = \frac{3}{8}$$

Notice that we *do not add* the '8's, because this 8 tells us how many equal pieces there are. The top numbers (1 and 2) tell us how many of these pieces (eighths) we have altogether (i.e. 3).

Similarly $\quad \frac{2}{5} + \frac{1}{5} = \frac{3}{5} \quad$ and $\quad \frac{3}{7} + \frac{2}{7} = \frac{5}{7}$

Now try these.

EXERCISE 10.6

1 $\frac{1}{4} + \frac{2}{4}$ **2** $\frac{2}{5} + \frac{2}{5}$

3 $\frac{1}{8} + \frac{4}{8}$ **4** $\frac{3}{10} + \frac{4}{10}$

5 $\frac{2}{6} + \frac{3}{6}$ **6** $\frac{1}{4} + \frac{1}{4}$

7 $\frac{2}{9} + \frac{5}{9}$ **8** $\frac{2}{12} + \frac{3}{12}$

9 $\frac{6}{100} + \frac{1}{100}$ **10** $\frac{3}{20} + \frac{5}{20} + \frac{2}{20}$

Example 7

At the party, $\frac{5}{8}$ of one cake has been eaten, and also $\frac{6}{8}$ of another. What fraction is this altogether?

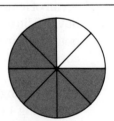

As usual, we can write

$$\frac{5}{8} + \frac{6}{8} = \frac{11}{8}$$

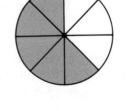

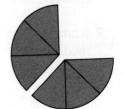

This is correct. However it is often useful to express a 'top-heavy' fraction as a whole number plus a fraction. It is the same as saying 'how many cakes can I make from $\frac{11}{8}$, and what is left over?'

If we take three of the six pieces and add them to the five pieces, we shall have one whole cake, and three pieces left over. We can say that

$$\tfrac{11}{8} = 1 + \tfrac{3}{8} \quad \text{or} \quad 1\tfrac{3}{8}$$

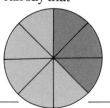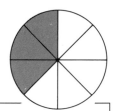

EXERCISE **10.7**

In these questions give your answer as a whole number plus a fraction.

1 $\tfrac{4}{7} + \tfrac{4}{7}$ **2** $\tfrac{2}{3} + \tfrac{2}{3}$

3 $\tfrac{7}{8} + \tfrac{4}{8}$ **4** $\tfrac{5}{6} + \tfrac{3}{6}$

5 $\tfrac{7}{10} + \tfrac{4}{10}$ **6** $\tfrac{8}{9} + \tfrac{7}{9}$

7 $\tfrac{3}{4} + \tfrac{2}{4}$ **8** $\tfrac{3}{5} + \tfrac{4}{5}$

9 $\tfrac{4}{5} + \tfrac{4}{5} + \tfrac{4}{5}$ **10** $\tfrac{7}{10} + \tfrac{8}{10} + \tfrac{9}{10}$

Going back to the party and the cake, Carol eats a piece. What fraction of the cake has now been eaten?

Altogether there will have been

$$\tfrac{1}{8} + \tfrac{2}{8} + \tfrac{1}{8} = \tfrac{4}{8}$$

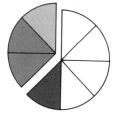

Now you can easily see that these four eighths are the same as half of the cake. We can write

$$\tfrac{4}{8} = \tfrac{1}{2}$$

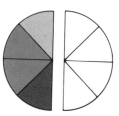

Similarly, the two pieces Mike ate came to a quarter of the cake. So we can write

$$\tfrac{2}{8} = \tfrac{1}{4}$$

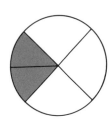

These **equivalent** fractions enable us to add (and subtract) fractions when cakes have been divided into different numbers of pieces.

Example 8

One cake is cut into quarters, and another into eight pieces. Harry takes a piece from each cake. Altogether, what fraction of one cake would these two pieces come to?

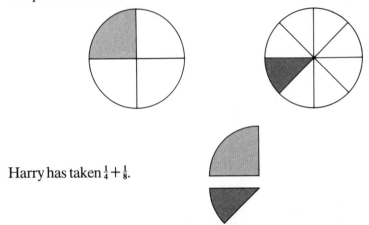

Harry has taken $\frac{1}{4} + \frac{1}{8}$.

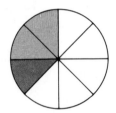

But we have seen, from Mike, that $\frac{1}{4}$ is the same as $\frac{2}{8}$.

So we can say that Harry has taken $\frac{2}{8} + \frac{1}{8} = \frac{3}{8}$.

We can add fractions *only* if the *bottom* numbers (**denominators**) are the *same*, as they were in Exercise 10.6.

The key to most work with fractions is to be familiar with equivalent fractions. We have met some already, with the cakes.

$$\frac{4}{8} = \frac{1}{2} \quad \text{and} \quad \frac{2}{8} = \frac{1}{4}$$

Here are some more:

$$\frac{4}{6} = \frac{2}{3} \quad \frac{10}{20} = \frac{1}{2} \quad \frac{6}{8} = \frac{3}{4} \quad \frac{9}{15} = \frac{3}{5} \quad \frac{1}{2} = \frac{6}{12} \quad \frac{5}{20} = \frac{1}{4}$$

If we want to find a small equivalent fraction (take $\frac{16}{20}$, for example) we need to find a number that will divide exactly into *both* the top number and the bottom number (either 2 or 4 in this case). Then do the division. In this case, $\frac{16}{20}$ will become $\frac{8}{10}$ or $\frac{4}{5}$.

EXERCISE 10.8

In these questions, work out the missing number in the equivalent fraction.

1 $\frac{5}{10} = \frac{}{2}$

2 $\frac{6}{8} = \frac{}{4}$

3 $\frac{2}{6} = \frac{}{3}$

4 $\frac{8}{12} = \frac{}{3}$

5 $\frac{9}{12} = \frac{}{4}$

6 $\frac{6}{10} = \frac{}{5}$

7 $\frac{12}{20} = \frac{}{5}$

8 $\frac{5}{15} = \frac{}{3}$

9 $\frac{6}{9} = \frac{}{3}$

10 $\frac{50}{100} = \frac{}{10} = \frac{}{2}$

It is often useful to find the equivalent fraction which you cannot make any simpler, because nothing else will divide exactly into both the top and the bottom of the fraction.

For example, $\frac{8}{12}$ can be simplified to $\frac{4}{6}$, and this in turn can be simplified again into $\frac{2}{3}$. We can't go any further, because there isn't a whole number which will go exactly into *both* 2 and 3 (except 1, which will divide into *all* whole numbers!). So $\frac{2}{3}$ is the simplest fraction **equivalent** to $\frac{8}{12}$.

EXERCISE 10.9

Try to find the simplest equivalent fraction in these questions.

1 $\frac{10}{20}$

2 $\frac{12}{18}$

3 $\frac{6}{9}$

4 $\frac{4}{14}$

5 $\frac{20}{50}$

6 $\frac{16}{40}$

7 $\frac{9}{36}$

8 $\frac{6}{12}$

9 $\frac{24}{40}$

10 $\frac{60}{100}$

Rather than simplifying fractions, there are times when we need to 'complicate' them, particularly when adding or subtracting.

Look back to the example where Harry took $\frac{1}{4}$ of one cake and $\frac{1}{8}$ of a second cake. Together, this was $\frac{1}{4} + \frac{1}{8}$. We 'complicated' the $\frac{1}{4}$ by doubling each number to give us the equivalent fraction $\frac{2}{8}$. This allowed us to add the fractions because we were now adding the *same* sized pieces.

$$\frac{2}{8} + \frac{1}{8} = \frac{3}{8}$$

EXERCISE **10.10**

Find the missing number in these equivalent fractions, by *multiplying* rather than dividing.

1 $\frac{3}{5} = \frac{}{20}$ **2** $\frac{1}{2} = \frac{}{16}$ **3** $\frac{2}{3} = \frac{}{12}$

4 $\frac{5}{8} = \frac{}{16}$ **5** $\frac{3}{10} = \frac{}{100}$ **6** $\frac{2}{7} = \frac{}{21}$

7 $\frac{4}{5} = \frac{}{10}$ **8** $\frac{1}{4} = \frac{}{}$ **9** $\frac{2}{3} = \frac{12}{}$

10 $\frac{8}{9} = \frac{40}{}$

Example 9

Add together $\frac{3}{5}$ and $\frac{1}{10}$.

How can I make the bottom numbers the same? It is quite easy to see that I can 'complicate' $\frac{3}{5}$ into $\frac{6}{10}$. So

$$\frac{3}{5} + \frac{1}{10} = \frac{6}{10} + \frac{1}{10} = \frac{7}{10}$$

EXERCISE **10.11**

Add these fractions. Remember to make the bottom numbers the same first, then add the top numbers.

1 $\frac{1}{2} + \frac{1}{4}$ **2** $\frac{3}{4} + \frac{1}{8}$

3 $\frac{2}{5} + \frac{3}{10}$ **4** $\frac{1}{6} + \frac{1}{2}$

5 $\frac{1}{3} + \frac{5}{12}$ **6** $\frac{1}{3} + \frac{1}{6}$

7 $\frac{3}{4} + \frac{7}{8}$ **8** $\frac{1}{2} + \frac{3}{10}$

9 $\frac{3}{8} + \frac{7}{24}$ **10** $\frac{3}{20} + \frac{1}{4}$

Subtracting fractions

Exactly the same principle applies to subtracting as to adding fractions. First make the bottom numbers the same (by equivalent fraction method) and then take away the top numbers.

Example 10

Work out $\frac{5}{6} - \frac{1}{3}$.

By making the $\frac{1}{3}$ into $\frac{2}{6}$, we can write

$$\frac{5}{6} - \frac{1}{3} = \frac{5}{6} - \frac{2}{6} = \frac{3}{6} \quad (= \frac{1}{2}, \text{ if we want to simplify any further})$$

EXERCISE 10.12

1 $\frac{1}{2} - \frac{1}{4}$ 2 $\frac{3}{4} - \frac{1}{8}$

3 $\frac{2}{3} - \frac{1}{6}$ 4 $\frac{7}{12} - \frac{1}{2}$

5 $\frac{9}{10} - \frac{1}{5}$ 6 $\frac{3}{8} - \frac{1}{4}$

7 $\frac{13}{20} - \frac{2}{5}$ 8 $\frac{1}{2} - \frac{1}{8}$

9 $\frac{7}{9} - \frac{1}{3}$ 10 $\frac{11}{15} - \frac{1}{3}$

Investigation B

(*a*) Use your calculator to work out $\frac{1}{7}$ as a decimal. You should get
0.142 857 1....

(*b*) Now work out $\frac{1}{7}$ by division – continue until you have at least 20
decimal places.

(*c*) You should have found that the figures 142857 keep repeating.

(*d*) Now work out $\frac{2}{7}$ by both methods. What do you notice?

(*e*) This time try $\frac{3}{7}, \frac{4}{7}, \frac{5}{7}$ and $\frac{6}{7}$. Try to remember the sequence of six numbers.

Extension

Repeat the process, for the fraction $\frac{1}{13}$.

Estimation

'About how many miles will a full tank of petrol take me in my car?'
'How many tiles, roughly, will I need for that wall?'
'How long is it going to take me to read that book?'

All of these questions, and many others, do not need an exact answer; an
estimate will be good enough for most purposes, and can usually be
worked out quite quickly in your head.

Example 11

I intend to drive from London to Carlisle. My road atlas gives the distance as 301 miles. If my car can normally travel for about 10 miles on one litre of petrol, about how many litres will it take for the journey?

Calling the distance 300 miles, it will take $\frac{300}{10} = 30$ litres approximately.

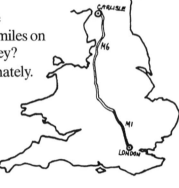

Example 12

It took me 28 seconds to read a page of a book. About how long will it take me to read the whole chapter of 24 pages?

Taking the 28 seconds as half a minute, it will take about half of 24, which is 12 minutes to read the chapter.

EXERCISE 10.13

1 A leg of pork, weighing 2.05 kg, costs £6.09. How much does the pork cost per kilogram, to the nearest 10p?

2 In a test match a batsman took 4 hours 56 minutes to score a century. About how many runs per hour did he score?

3 The bill for the meal I had at a hotel came to £7.89. Next week I intend to book the same hotel for a meal for six people. About how much will I need to take to cover the cost of the six meals?

4 One innnings of a cricket match lasts 42 overs. Our side takes about three and half minutes to bowl one over. How long is this innings likely to last, to the nearest quarter of an hour?

5 After checking my gas meter, I find that I have used 152 therms since paying the last bill. If gas costs 41.5p per therm, about how much will this gas cost me, to the nearest £10?

6 About how many tiles, 20 cm square, will I need to cover a wall measuring 118 cm by 77 cm? (Squared paper might help, and assume that you don't break any!)

7 One turn of the reel extends my fishing line by 10.3 cm. If it takes 292 turns to extend my line fully, about how long is my line? Give your answer in metres.

8 About how long will it take to cook a chicken weighing 1.93 kg, if it takes 20 minutes for each 500 grams?

9 A television advert lasts for 23 seconds. The company agrees to show the advert 16 times in the next week. About how many minutes of advertising will the firm have to pay for?

10 A cooker costs £399. I agree to pay for it over two years, paying 24 equal instalments. If the hire purchase charge comes to £73.68, about how much will each instalment be, to the nearest £?

11. Measuring: graphics

Reading scales and dials

Scales and dials take many forms. The easiest to read is the digital kind where the reading is presented as a series of numbers.

1379840

Calculators, digital watches and numerical meters use this type of display. They are becoming increasingly common.

Meters can sometimes use a series of dials to represent readings, each dial representing a digit of the reading.

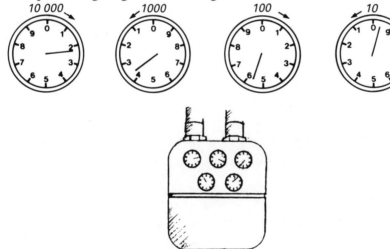

To read each dial you will normally find the needle is between two numbers: take the lowest number as the reading, except between 0 and 9 when 9 should be read. The reading above gives us 23591.

Clocks and watches can be either digital or with a traditional dial. Compare the times on these watches – they tell us the same time in a different way.

EXERCISE 11.1

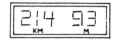

1 (*a*) How many copies have been made as recorded on the printer reading?

(*b*) What would be the reading if 1000 more copies were made?

2 (*a*) What is the reading on the exercise bike?

(*b*) Write down the reading after a further 10 metres has been added.

(*c*) Another bike has a dial showing km only as a decimal. How would this second bike show both the readings from the first bike?

3 (*a*) Write down the temperature as shown.

(*b*) If this is the temperature of a room would it be in °F or °C?

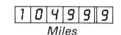

Temperature

4 (*a*) How many miles does the reading show?

(*b*) What does the last digit tell you?

(*c*) What would be the next reading to appear on the dial?

Miles

5 (*a*) What number is shown by the video counter?

(*b*) How many times will the last digit change before the 1 becomes a 2?

Counter

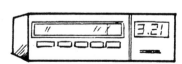

6 (*a*) What time is shown on the video clock?

(*b*) What time will it show after 5 minutes more?

VHS

Write down (*a*) the times as shown by the watches, (*b*) the times as they would be shown on a digital watch:

7   **8**   **9**   **10**

Write down the times as shown on the digital watches, and draw traditional style watches with dials to show the same time.

11  **12**  **13**  **14**

1:15 *11:50* *8:10* *00:05*

15 Describe the time as shown on the digital watch. *3:51* 23

EXERCISE 11.2

Write down the readings shown.

1

2

3

4

Draw dials to show the following readings.

5 24027   43210   14203   59875   67103   97427

Investigation A

The display shows how a calculator indicates numbers, which are made up of a series of 'bars'.

(*a*) Which of the above numbers have a value the same as their number of bars?

(*b*) What is the maximum number of bars which can be lit on a calculator with a display which can show a maximum of eight numbers, and what number will be shown?

(*c*) What number can be made using just 2 bars? 3 bars? Which numbers can be made using 4 bars?

(*d*) Using 5 bars we can make the numbers 71, 17, 3, 5, or 9, that is 5 numbers altogether. How many numbers can we make using 6, 7 and 8 bars? Investigate this idea further.

Decimals

Many readings involve the use of decimals. You will already be used to reading off decimal measurements from your ruler, so you could also do the same for dials.

Estimation

Sometimes the dials do not show accurately the readings required, and an estimate of the reading has to be taken.

Is this 2.2, 2.3, or 2.4? You will need to estimate the distance moved by the needle.

Is this reading 2.33 or 2.34?

EXERCISE 11.3

Write down the readings in each of the diagrams. For some you will
have to estimate a reading.

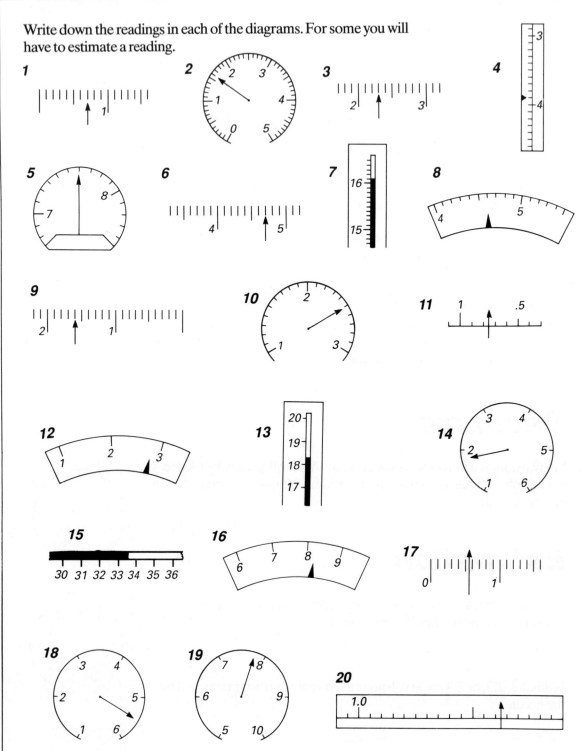

Conversion graphs

Both imperial and metric units will be used for a long time to come in this country, and occasionally we want to change between imperial and metric units. There are various methods available to make this easier for us, including conversion graphs.

Example 1

We can use the graph to read off a conversion.

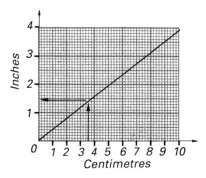

To change 3.6 cm we use the graph as shown to give us 1.42 inches.

EXERCISE 11.4

1 (*a*) Change into km (i) 2 miles (ii) 3.5 miles (iii) 4.3 miles.

   (*b*) Change into miles (i) 4 km (ii) 6.5 km (iii) 7.2 km.

2 (*a*) Change into kg (i) 2 lb (ii) 3½ lb (iii) 5¼ lb.

   (*b*) Change into lb (i) 2.5 kg (ii) 0.8 kg.

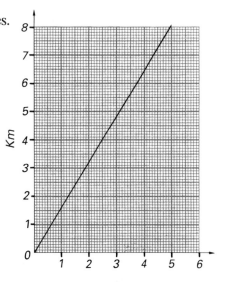

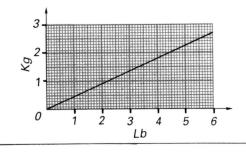

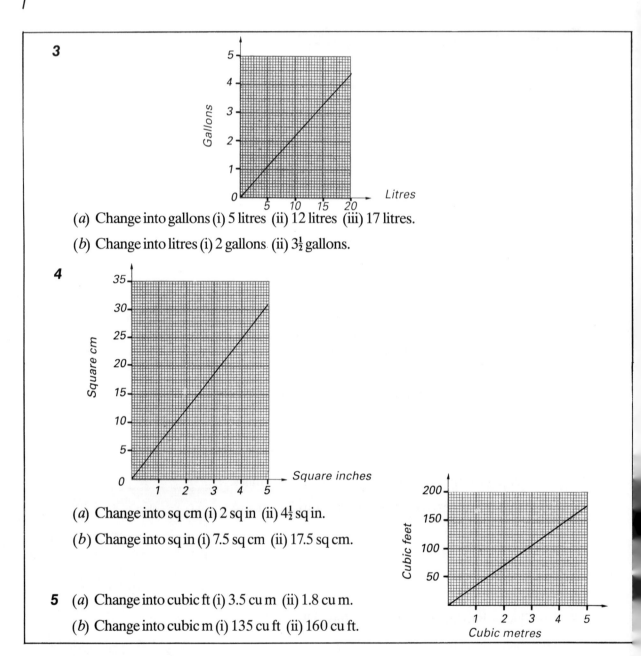

3

(*a*) Change into gallons (i) 5 litres (ii) 12 litres (iii) 17 litres.

(*b*) Change into litres (i) 2 gallons (ii) $3\frac{1}{2}$ gallons.

4

(*a*) Change into sq cm (i) 2 sq in (ii) $4\frac{1}{2}$ sq in.

(*b*) Change into sq in (i) 7.5 sq cm (ii) 17.5 sq cm.

5 (*a*) Change into cubic ft (i) 3.5 cu m (ii) 1.8 cu m.

(*b*) Change into cubic m (i) 135 cu ft (ii) 160 cu ft.

Investigation B

Many packets or containers in your house or in the shops show
measurements written in both imperial and metric. Make a note of these
measurements, and then plot them on graph paper to produce a
conversion graph; grams to ounces, kg to lb, pints to litres. How accurate
are the conversions used on the packets?

Conversion charts

Example 2

Some rulers can be used as a chart to help convert centimetres to inches.

To change 4 cm to inches we can read the scales and estimate it to be about 1.6 inches. Alternatively 2.5 inches = 6.2 cm.

EXERCISE 11.5

1 Using the scale on the measuring jug,

(*a*) change the following into fluid ounces (i) 0.1 litres (ii) 0.25 litres (iii) 0.17 litres;

(*b*) change the following into litres (i) 5 fl.oz (ii) $7\frac{1}{2}$ fl.oz (iii) 3 fl.oz.

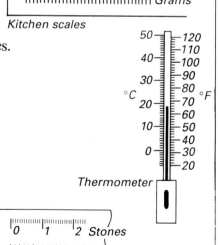

Measuring jug

2 Using the readings from the kitchen scales,

(*a*) change into lb and ounces (i) 500 g (ii) 220 g (iii) 350 g;

(*b*) change into grams (i) 12 oz (ii) 9 oz (iii) 1 lb 4 oz.

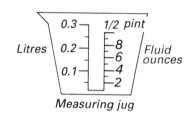

Kitchen scales

3 The thermometer can help you to change these temperatures.

(*a*) Change into °C (i) 70°F (ii) 26°F (iii) 102°F.

(*b*) Change into °F (i) 35°C (ii) −2°C (iii) 28°C.

4 The first part of the dial of some bathroom scales can help convert stones and pounds into kilograms.

(*a*) From the dial can you tell how many lbs make a stone?

(*b*) Change into st and lb (i) 5 kg (ii) 12 kg (iii) 16.5 kg.

(*c*) Change into kg (i) 1 st 4lb (ii) 6 lb (iii) 1 st 10 lb.

Thermometer

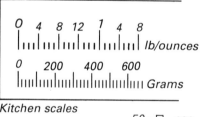

Bathroom scales

Example 3

1 inch = 2.54 cm 2 inches = 5.08 cm 3 inches = 7.62 cm 4 inches = 10.16 cm 5 inches = 12.7 cm 8 inches = 20.32 cm 10 inches = 25.4 cm 15 inches = 38.1 cm 20 inches = 50.8 cm	To convert a measurement such as 18 inches, we can do it by *adding* the amounts like this: 18 inches = 15 inches + 3 inches = 38.1 cm + 7.62 cm = 45.72 cm

EXERCISE 11.6

1   Change into millilitres (i) 6 fl. oz (ii) 8 fl. oz (iii) 16 fl. oz.

2   Change into litres (i) 5 pints (ii) 7 pints (iii) 9 pints.

3   Change into grams (i) $2\frac{1}{2}$ oz (ii) 6 oz (iii) $12\frac{1}{2}$ oz.

1 fl oz = 28 ml 2 fl oz = 57 ml 3 fl oz = 85 ml 4 fl oz = 114 ml 5 fl oz = 142 ml 10 fl oz = 284 ml	1 pint = 0.568 l 2 pint = 1.137 l 3 pint = 1.705 l 4 pint = 2.273 l 6 pint = 3.408 l	$\frac{1}{2}$ oz = 14g 1 oz = 28g 2 oz = 57g 3 oz = 85g 4 oz = 113g 8 oz = 227g

Investigation C

We need a graph to find out the relationship between shoe sizes and sizes of feet. Conduct a survey among the members of your class: measure their feet in centimetres, and record the shoe size they are wearing. Once you have all your results you can then record the information on a graph, and attempt to draw in an approximate line to make a conversion graph between these two values.

Angles

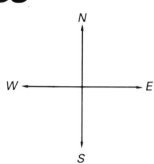 

The diagram shows the four points of the compass. The angle between each of the four compass points is a right angle, that is 90°. Use your protractor to draw the compass rose accurately into your book, taking about half a page to do it.

There are other points we need to add to our compass as shown. What will be the angle between North and North East?

North East is exactly half-way between North and East. The angle between North and East will be half a right angle, or 45°. All the angles on this compass are 45°. Use your protractor to add the other four points to the compass you have already drawn.

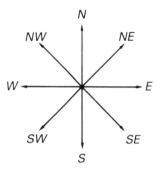

EXERCISE 11.7

Find out how many degrees there are between these points of the compass measured clockwise.

1 N and S.   **2** N and W.   **3** S and E.   **4** W and E.   **5** N and NE.

6 E and SW.   **7** NW and E.   **8** N and NW.   **9** W and S.   **10** SW and SE.

11 If you were facing SW and turned 45° clockwise, in what direction would you be facing?

Find the direction in which you would be facing after making the following clockwise turns.

12 Facing SE and turning 180°.   **13** Facing NE and turning 135°.

14 Facing W and turning 270°.   **15** Facing SW and turning 90°.

16 Facing N and turning 315°.   **17** Facing S and turning 360°.

18 Facing E and turning 225°.   **19** Facing W and turning 135°.

20 Facing SE and turning 270°.

Investigation D

Further points can be added to help read the compass. What would you call the point between N and NE? If you are uncertain find out, then construct a compass with these extra points. The simplest compass had 4 points, the next 8 points. How many has your new compass? Could you go on to draw the next compass in the series using a larger sheet of paper?

Using protractors

Using protractors we can also measure angles.

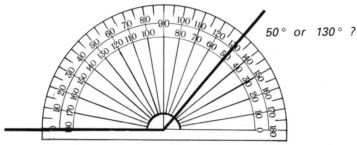

50° or 130° ?

When we use a protractor there are two readings. Which one do we use? Remember, any angle less than 90° is called an **acute** angle. Any angle between 90° and 180° is called an **obtuse** angle.

acute

obtuse

The angle in the second diagram is clearly *more* than 90°, and obtuse, so the correct reading must be 130°, and not 50° which is less than 90°.

EXERCISE 11.8

Measure the following angles.

1

2

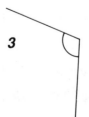

3

4

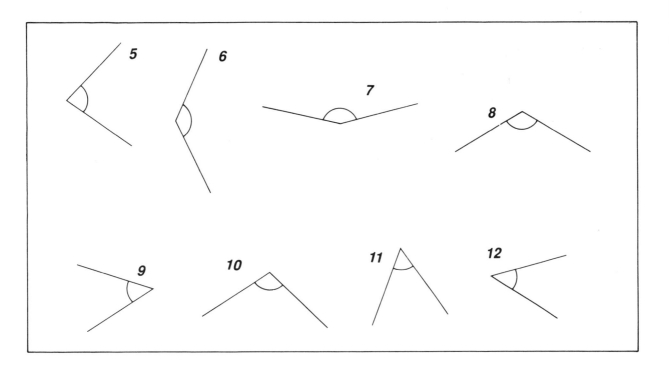

EXERCISE 11.9

Draw the following angles.

1 50°   **2** 75°   **3** 15°   **4** 165°   **5** 95°

6 125°   **7** 43°   **8** 142°   **9** 105°   **10** 65°

Measuring angles more than 180° is more difficult, but it can be done.

Measure the dotted angle first. We know that both angles add up to 360°, so to find the angle we want, take the measured dotted angle away from 360°:

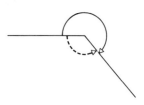

dotted angle = 130°
angle required = 360° − 130° = 230°

To draw angles more than 180° we use a similar method.

To draw the angle 245°, first calculate the other angle which goes with it: 360° − 245° = 115°. Now draw this angle (115°) and the angle we want will be drawn at the same time. Remember to clearly mark the angle 245° in your diagram, as this is the one you want.

EXERCISE 11.10

Measure the following angles.

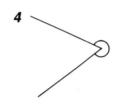

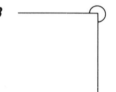

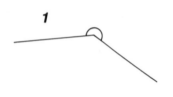

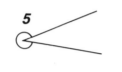

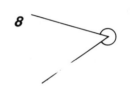

EXERCISE 11.11

Draw the following angles.

1 200°	**2** 280°	**3** 310°	**4** 345°	**5** 215°
6 325°	**7** 245°	**8** 300°	**9** 225°	**10** 355°

EXERCISE 11.12

Now measure these angles, but first identify whether they are less than or more than 180°.

1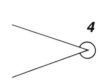

2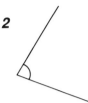

3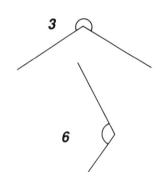

4

5

6

Draw the following angles.

7 60°   **8** 330°   **9** 135°   **10** 235°   **11** 150°

12 305°   **13** 25°   **14** 85°   **15** 225°

Bearings

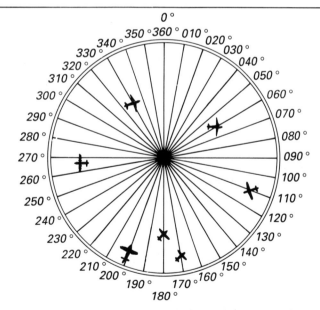

Bearings are directions given in terms of an angle. Notice from the radar in the diagram that bearings are always expressed as three-figure numbers. 10° is written as 010°. Angles are measured *clockwise* from North.

EXERCISE 11.13

1 Write down the bearing of each of the aircraft on the radar scope.

2 Find the bearing of each of the points on your compass rose: NE, E, SE, etc.

Measure the following bearings.

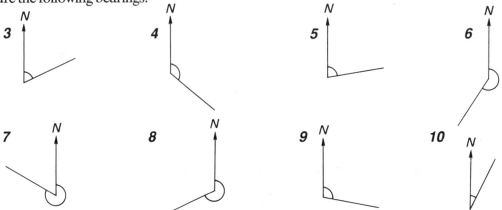

Draw the following bearings. Remember to start with a vertical North line for every question.

11 030°	**12** 120°	**13** 175°	**14** 210°	**15** 295°
16 340°	**17** 085°	**18** 155°	**19** 180°	**20** 270°

Investigation E

Stand in the centre of the playground. Using a compass, note down all the important or prominent things you see around you, and their direction from where you stand. Include corners of buildings, any trees, houses, gates, etc. You can then construct a radar map of what you have noted. You could also estimate or measure the distances of the points noted from where you were standing.

Using bearings

Bearings are used mainly in solving navigational problems. In this sort of problem we also need to measure distances to scale and you will need to remember the work covered in Chapter 6 on map scales.

Example 4

A pilot starts his journey by flying from Abingdon on a bearing of 050°
for 400 km to Baguley. He changes direction above Baguley to a bearing
of 120° and flies for a distance of 350 km to Cresham. Using a scale of
1 cm to 100 km, draw a scale drawing of his route.

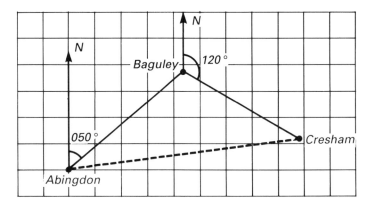

Notice when we draw a bearing at Abingdon and Baguley we first draw a
vertical North line. If you wanted to find the direct distance from
Abingdon to Cresham you can now measure it on the diagram as shown
by the dotted line, and convert the measured distance in cm back into km:

$$6.25 \text{ cm} \times 100 = 625 \text{ km}$$

To find the bearing of Cresham from Abingdon we measure the angle
(bearing) to the dotted line, as this is the line from Abingdon to Cresham.
You should find that the bearing is 082°.

EXERCISE 11.14

1 Copy and complete the sentence: 'The pilot took off from
Roslyn and headed towards Tarn, _____ km away on a
bearing of _____. After flying over Tarn he took a new
bearing of _____ and flew a further distance of
_____ km to Bleakley.'

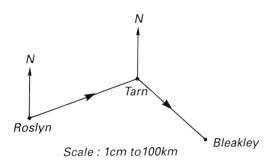

Scale : 1cm to 100km

2 Copy and complete the sentence: 'They flew from Torell towards Marks Isle, _____ km away on a bearing of _____. After flying over Marks Isle they took a new bearing of _____ and flew a further distance of _____ km to Eadale.'

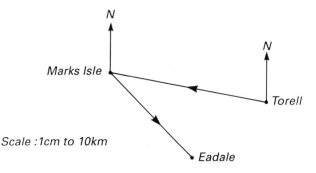

Scale : 1cm to 10km

3 Copy and complete the sentence: 'The pilot took off from Edendale and first headed towards Bardon, _____ km away on a bearing of _____. After flying over Bardon, he took a new bearing of _____ and flew a further distance of _____ km to Firswood, followed by a third bearing of _____ to take him the _____ km to Oakley. He then flew a distance of _____ km back to Edendale.'

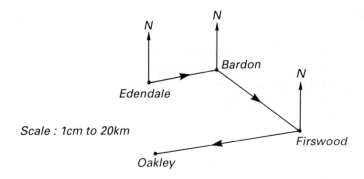

Scale : 1cm to 20km

EXERCISE 11.15

1 Describe the distance and bearing of each point from A.

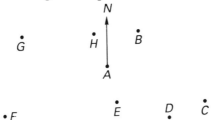

2 Make an accurate scale drawing of the sketch of the journey, using a scale of 1 cm to 10 km.

(*a*) How far is it in a straight line from Avenham to Chatburn?

(*b*) What is the bearing of Chatburn from Avenham?

(*c*) What is the bearing of Avenham from Chatburn?

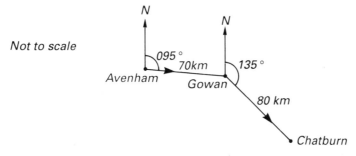

3 Make an accurate scale drawing of the sketch of the journey, using a scale of 1 cm to 100 km.

(*a*) What is the distance from Broadhill to Hawthorns?

(*b*) What is the bearing of Hawthorns from Alston?

(*c*) What is the distance and bearing of Broadhill from Alston?

(*d*) How far is it from Bramley to Hawthorns?

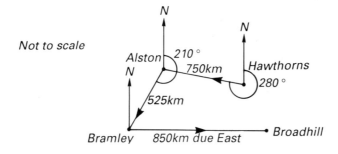

4 A boat takes a course bearing 160° for 600 km followed by a bearing of 050° for 700 km. Draw a scale diagram of its course, and find the bearing and distance if the journey was to be completed in one stage.

5 Draw a scale diagram of the following journey: 450 km on a bearing of 090°, followed by a bearing of 200° for 350 km.

6 A pilot is going to make a journey calling at three airports before returning home. After taking off he takes a bearing of 050°, and flies for 550 km. His next course is on a bearing of 150°, for 510 km. To get to the final airport requires a course bearing of 245° for 650 km. Make an accurate scale drawing of the journey planned.

   (*a*) What bearing will the pilot have to take to get home, and what distance will have to be flown?

   (*b*) Describe the journey in terms of bearings and distances if the journey to the three airports were to be repeated, but in reverse order.

Investigation F

'Treasure Hunt' is a game for small groups working either inside or outside. Each group designates a spot where its treasure is buried, and describes the route to be taken from the classroom door in terms of distances and directions (use compass points or bearings). The description is then passed on to another group which then has to find the route and arrive at the same spot, and the 'treasure'!

12. Measuring: speed, averages and rates

Average speed

Example 1

You ride your bike round a track in 50 seconds. If the track is 350 metres long, what is your average speed?

Your average speed is **total distance/total time**, which is

$$\frac{350}{50} = 7 \text{ metres per second}$$

Your speed will change all the time – your **average speed** is the steady speed you would need to ride in order to cover the 350 metres in the same time of 50 seconds, i.e. 7 metres per second.

In general we can say that

$$\textbf{Average speed} = \frac{\textbf{total distance covered}}{\textbf{total time taken}}$$

Example 2

A train leaves Newcastle at 7.30 a.m. and arrives at King's Cross, London, at 10.30 a.m., having travelled 438 kilometres. What is the average speed of the train in kilometres per hour?

Time taken is three hours (7.30 until 10.30).

$$\begin{aligned} \text{Average speed} &= \frac{438}{3} \\ &= 146 \text{ km/h} \end{aligned}$$

EXERCISE 12.1

1 In two hours I jogged 30 kilometres. What was my average speed?

2 An aeroplane flies 1500 km in four hours. Work out its average speed.

3 If it takes me three hours to drive the 141 miles from Cardiff to Portsmouth, what is my average speed, in miles per hour?

4 A sprinter runs 200 m in 25 seconds. What is her average speed, in metres per second?

5 It takes me 7 minutes to walk from my house to the youth club. I estimate the distance to be 840 metres. What is my average walking speed, in metres per minute? How many metres per second is this?

6 Robert is going to visit his aunt, who lives 140 miles away. The coach leaves at 9.30 a.m. and arrives at his aunt's house at 1.30 p.m. What is the average speed of the coach, in miles per hour?

7 An inter-city train leaves London at 1615 and is due to arrive in Plymouth at 1845. Work out the average speed of the train, if the distance from London to Plymouth is 210 miles.

8 In a race a runner covers the first 400 m in 50 seconds. What was his average speed, in metres per second?

9 In January 1957 an American B-52 aeroplane flew round the earth in 45.32 hours. It travelled a distance of 24 325 miles. What was its average speed, in miles per hour?

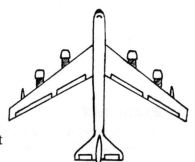

10 A jogger takes $1\frac{1}{2}$ hours to cover a distance of 18 km. Work out his average speed in kilometres per hour.

Averages

Example 3

Three boxes of matches contain 46, 48 and 53 matches. What is the average number of matches in a box?

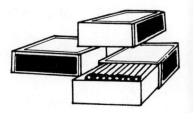

Average number $= \dfrac{\text{total number of matches}}{\text{total number of boxes}}$

$$= \dfrac{46 + 48 + 53}{3}$$

$$= \dfrac{147}{3}$$

$$= 49 \quad \text{matches per box}$$

Example 4

In his last five innings a batsman scored 17, 52, 0, 13 and 33 runs. What is his batting average, i.e. the average number of runs scored per innings?

Average number of runs per innings

$$= \dfrac{\text{total number of runs}}{\text{total number of innings}}$$

$$= \dfrac{17 + 52 + 0 + 13 + 33}{5}$$

$$= \dfrac{115}{5}$$

$$= 23 \quad \text{runs per innings}$$

EXERCISE 12.2

1 A packet of the same brand of washing powder, in four different shops, costs £1.37, £1.25, £1.29 and £1.29. What is the average cost?

2 Six men weigh 83.1 kg, 84 kg, 83.7 kg, 90 kg, 82.5 kg and 84.3 kg. What is their average weight?

3 The times it took me to travel to school on each day last week were 23, 25, 25, 31 and 26 minutes. What was my average travelling time?

4 Three of my classmates do a paper round. Last week their pay was £7.21, £9.47 and £7.41. What was their average weekly pay?

5 While doing a project, I counted the number of sweets in ten packets. The totals were 27, 31, 32, 25, 25, 30, 27, 34, 32 and 27. What was the average number of sweets in a packet?

6 My electricity bills for the four quarters of last year were
£121.73, £109.68, £84.91 and £67.08. If I had decided to pay
my bills by twelve monthly payments, what would the average
monthly payment have been?

7 In marking the work of eight pupils in a drawing test, the teacher
counted the number of mistakes as 5, 12, 3, 0, 16, 2, 7 and 3.
What was the average number of mistakes per pupil?

8 My mother grew nine tomato plants last year, and the numbers
of edible tomatoes we picked from these plants were 18, 16, 18,
25, 13, 16, 3, 0 and 17. What was the average number of edible
tomatoes per plant?

9 The first train in the morning takes 38 minutes from the station
to town. The next three trains take 39, 43 and 31 minutes. What
is the average time taken by these four trains for this journey?

10 The weights, in kilograms, of the eight men in the pack of
forwards in our rugby team are 98, 103, 94, 97, 110, 104, 86 and
98. Work out their average weight.

Speed: distances and times

Example 5

I expect to be able to maintain an average speed of 50 miles per hour on a
car journey. How far will I expect to travel in three hours?

As I expect to travel 50 miles in each hour, I expect to travel a distance of
$50 \times 3 = 150$ miles.

If you know the *speed* and the *time taken* at that speed, then the

   distance travelled = speed × time

Example 6

A racing car averages 150 kilometres per hour. How far will it travel in a
race lasting $2\frac{1}{2}$ hours?

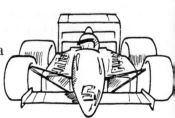

Distance = speed × time = $150 \times 2\frac{1}{2}$ = 375 kilometres.

Example 7

If I can ride my bike at an average speed of 15 kilometres per hour, how far will I go in 20 minutes?

Care is needed here, because the speed is in kilometres per *hour*, whereas the time taken is given in *minutes*.

(*a*) *working in hours:* 20 minutes $= \frac{1}{3}$ of an hour.
Distance that you could travel $=$ speed $\times$ time
$$= 15 \times \tfrac{1}{3}$$
$$= 5 \text{ kilometres}$$

(*b*) *working in minutes:* 15 km/h $= \frac{15}{60}$ kilometres per minute
$$(\, = \tfrac{1}{4} \text{ kilometre per minute})$$

Distance that you could travel $= \dfrac{15}{60} \times 20$
$$= \tfrac{1}{4} \times 20$$
$$= 5 \text{ kilometres}$$

EXERCISE 12.3

1 How far, at 4 metres per second, can you travel in 20 seconds?

2 I drive for four hours to visit my sister's new house. If I drive at 40 miles per hour on average, how far away is my sister's new house?

3 In a race from one end of the village to the other, the winner's time was 6 minutes. If the winner was running at an average speed of 20 kilometres per hour, how long is the village?

4 A passenger ferry, averaging 15 km/h, crosses the English Channel in $2\frac{1}{2}$ hours. How far is the crossing?

5 Walking at 6 km/h, I can get home from the station in 15 minutes. How far is the station from my home?

6 In a motorcycle race it took a rider 4 minutes to complete one lap, at an average speed of 200 km/h. How far is one lap?

7 In an experiment a metal block was pulled at a steady speed of 2.3 metres per second. How far will it move in (*a*) 3 seconds, (*b*) 5.3 seconds?

8 A '125' train travels at 108 miles per hour. How far will it travel in (*a*) half an hour, (*b*) 20 minutes, (*c*) 5 minutes?

9 A cheetah can run at a speed of 90 km per hour. What distance could it cover in (*a*) 10 minutes, (*b*) 1 minute?

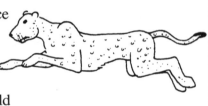

10 A competition to win a car involves guessing how far it can travel on a full tank of petrol, at a steady speed of 60 km/h. If you estimate that it will travel for $7\frac{1}{2}$ hours, what distance would you expect it to travel?

Example 8

How long will it take a train to cover a distance of 270 km, travelling at 90 km/h?

If it takes an hour to travel 90 kilometres, we need to find out how many 90s there are in 270.

$$\text{Time taken} = \frac{270}{90} = 3 \text{ hours}$$

If you know the *speed* and the *distance* travelled, then

$$\textbf{time taken} = \frac{\textbf{distance}}{\textbf{speed}}$$

EXERCISE 12.4

1 How long will it take a car, travelling at an average speed of 45 miles per hour, to travel a distance of 180 miles?

2 An aeroplane averages 500 km/h when flying a distance of 3000 km. How many hours does the flight last?

3 One lap of a model car racing track is 35 metres long. If my best car does 5 m/s, how long will the car take to complete a lap of the racing track?

4 How long will it take a runner, running at 8 metres per second, to run 400 metres?

5 A greyhound covered a distance of 480 metres at an average speed of 15 m/s. How many seconds did the greyhound take?

6 A motor race is over 810 kilometres. If a racing car can average 180 km/h, how many hours will it take to complete the race?

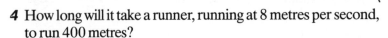

7 How long will it take a train travelling at an average speed of 70 miles per hour to cover the 140 miles from Crewe to Watford Junction?

8 My walking speed is 6 kilometres per hour. How long will it take me to walk (*a*) 12 km, (*b*) 2 km (*c*) 20 km?

9 The Indianapolis 500 race, a distance of 800 km, was won at an average speed of 265 km/h. How long did it take the winner? (Give your answer to the nearest minute.)

10 At 35 metres per second, how long will it take to travel 2.1 km?

13. Measuring: shapes

Perimeter

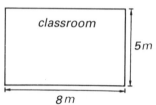

classroom

5 m

8 m

The **perimeter** of a shape is the distance all the way round it.

If we wanted to walk around the four walls of the classroom we would walk the following distance: 5 m + 8 m + 5 m + 8 m, so the perimeter of the classroom is 5 m + 8 m + 5 m + 8 m = 26 m.

Example 1
The table needs a protective plastic strip fastening around the edge of the table. What length of strip is needed?

$$2 \, m + 1 \, m + 2 \, m + 1 \, m = 6 \, m \quad \text{of strip.}$$

1 m

2 m

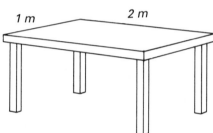

EXERCISE 13.1

Find the perimeter of each of the shapes below.

1
4 m

6 m

2
4 m

2 m

3
5 m   5 m

5 m

4
3 cm

3 cm

5
50 ft

15 ft

6
2.5 cm

2.5 cm

7
2.5 cm   3 cm

3 cm

6 cm

8
6 cm

4 m   2 m

1 m

2 m   2 m

3 cm

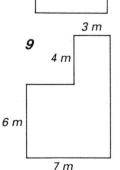

9
3 m

4 m

6 m

7 m

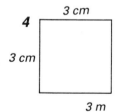

10
3 m   3 m

4 m   2 m

10 m

2 m

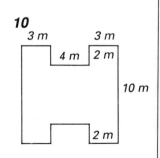

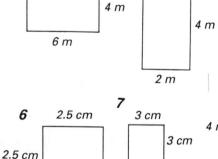

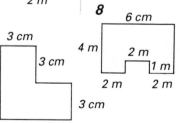

EXERCISE **13.2**

Find the perimeter of each of the shapes below.

1 A rectangle of length 8 m and width 4.5 m.

2 A square with each of its sides 5 cm long.

3    **4**   **5**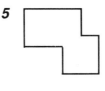

6 A field is fenced in and the fence posts are placed 1 m apart. What is the perimeter of the field?

7 The perimeter of the rectangle is 22 m. What is its width?

6 m

8 The perimeter of this rectangular steel plate is 36 cm. Find the width.

12 m

9 The perimeter of this piece of card is 21 cm. What is its length?

4 cm

10 The perimeter of this square piece of wood is 32 cm. Find the length of one of its sides.

Investigation A

To find the perimeter of a square or a rectangle we can write down the length of the four sides, and add them up. Can you describe a quicker way of finding the perimeter for (*a*) a square, and (*b*) a rectangle? It may help you to draw some squares and rectangles and work out the perimeters.

Area

Area is a measure of surface, or the measure of the amount of surface contained within a shape.

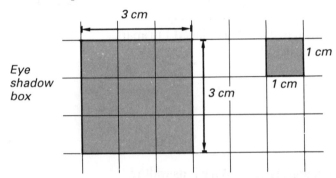

The centimetre square is the unit we use to measure the area of the eye shadow box. We want to know how many centimetre squares (cm^2) are contained within the eye shadow box. This will then be a measure of its area. There are 9 squares inside, so the area is 9 cm^2.

Alternatively, the eye shadow box is 3 cm long and 3 cm wide; i.e. 3 by 3. The area is then 3 cm $\times$ 3 cm = 9 cm^2.

Example 2

Find the area of the comb case.

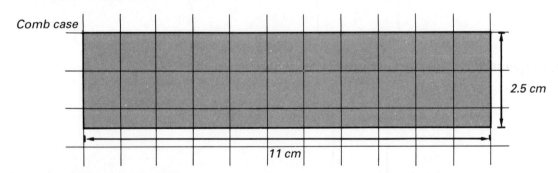

Count the squares inside the comb case to find the area. Don't forget to add up the half squares as well.

Alternatively, by multiplication, the comb case is 11 cm long and $2\frac{1}{2}$ cm wide. The area is therefore 11 cm $\times$ $2\frac{1}{2}$ cm = $27\frac{1}{2}$ cm^2. Is this how many squares you counted?

Investigation B

On a sheet of either squared paper or graph paper draw a rectangle, and count the squares to find the area. Now measure the length and width, and multiply them together. Compare your two answers; they should be the same. Now try this exercise with some other rectangles.

EXERCISE 13.3

Find the area of the following shapes.

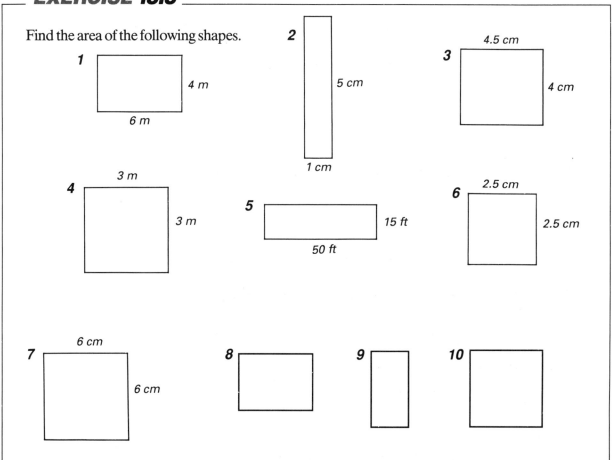

EXERCISE 13.4

1 Find the area of a rectangular piece of fabric with length 14 cm and width 12 cm.

2 Find the area of the floor of a square office with sides of length 8.4 m.

3   The area of this rectangle is 20 cm². Find the width.

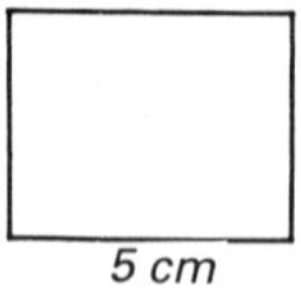

5 cm

4   The area of this rectangle is 50 m². What is the length?

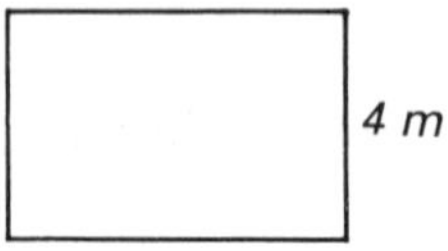

4 m

5   The area of this rectangle is 50 mm². Find the width.

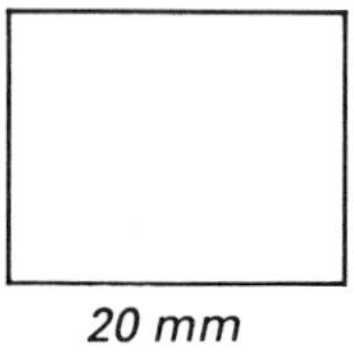

20 mm

6   The area of the square is 16 cm². What is the length of each side?

7   The area of a square is 81 m². What is the length of each side?

In questions 8 – 10 find out which *two* shapes in each question have the same area.

8

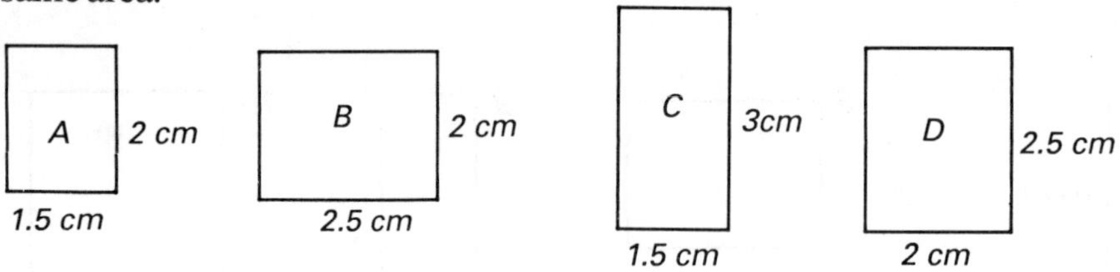

A *2 cm*
1.5 cm

B *2 cm*
2.5 cm

C *3cm*
1.5 cm

D *2.5 cm*
2 cm

9

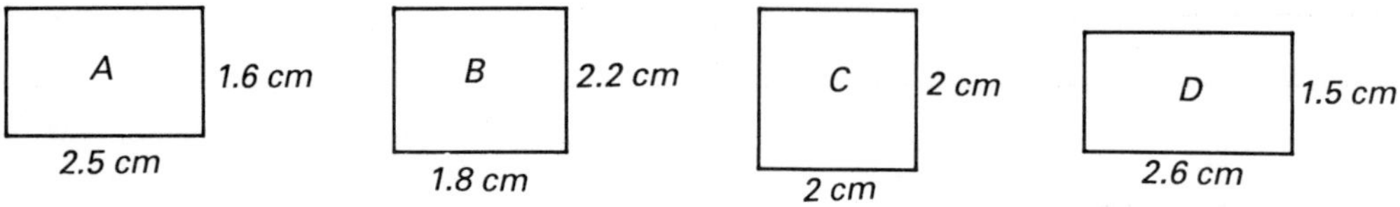

A *1.6 cm*
2.5 cm

B *2.2 cm*
1.8 cm

C *2 cm*
2 cm

D *1.5 cm*
2.6 cm

10

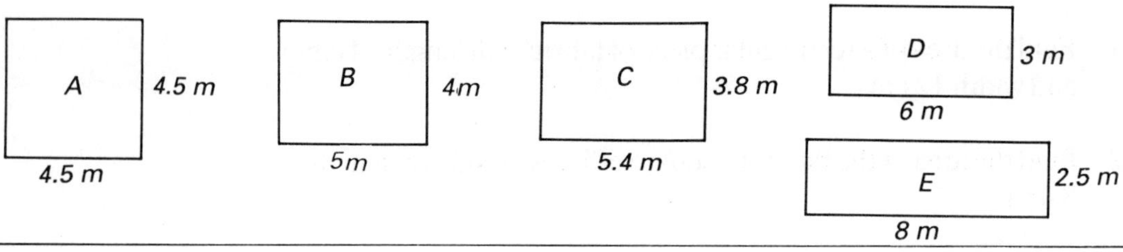

A *4.5 m*
4.5 m

B *4m*
5m

C *3.8 m*
5.4 m

D *3 m*
6 m

E *2.5 m*
8 m

Investigation C

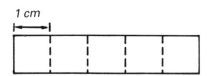

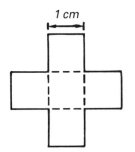

These shapes have a perimeter of 12 cm. How many other different shapes can you find which have a perimeter of 12 cm? Which of these shapes have a maximum area? Investigate other shapes with larger perimeters.

Investigation D

A farmer has enough fencing to build a pen of perimeter 40 m for his sheep.

(*a*) Write down a list of measurements, in terms of lengths and widths, he could use to put up his fence to make a four-sided pen.

(*b*) Calculate the area inside each of the pens you have described.

(*c*) Can you find the measurements which would give a maximum area for the pen, given a perimeter of 40 m?

Area of floors

In laying carpets you have to fit a carpet to match the room, and sometimes the shape of the room is difficult to measure.

Example 3

Find the area of carpet to be laid (shaded area).

Area of the whole rectangle $= 10 \times 5 = 50 \, \text{m}^2$
Area of piece to be cut away $= 4 \times 2 = \underline{8 \, \text{m}^2}$
Area of carpet to be laid $ = 42 \, \text{m}^2$

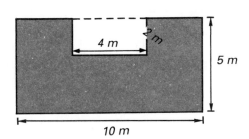

Example 4

Find the area of the border (shaded area).

Area of floor $= 12 \times 6 = 72\,m^2$
Area of carpet $= 10 \times 4 = \underline{40\,m^2}$
Area of border $= 32\,m^2$

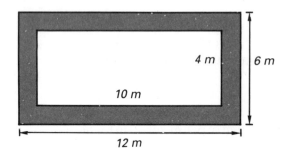

EXERCISE 13.5

In each case calculate the area of the shaded portion.

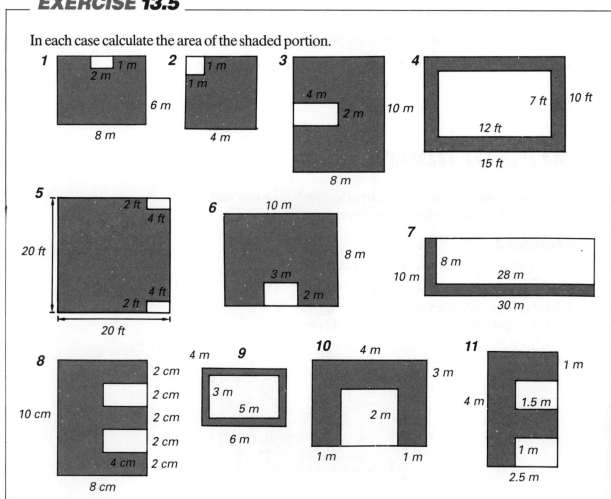

Investigation E

Jeremy wants to build an artificial fireplace in his room, but is not sure where to put it. Will its position affect the perimeter of the room? Draw several plans of the room with the fireplace in different positions and calculate the perimeter in each case. Does it change? Can you explain your result?

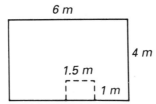
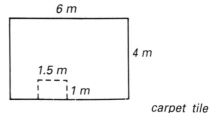

Investigation F

The floor of a room is to be carpeted using the tiles as shown.

(*a*) Draw a scale diagram and find out how many carpet tiles are needed.

(*b*) Find out how many of each tile will be needed if two colours of tile are used, to be laid in the pattern shown.

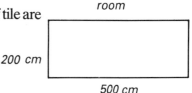

Area of a triangle

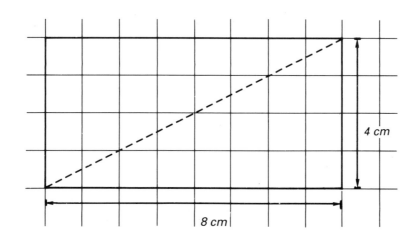

The area of the rectangle is 32 cm². Each triangle has an area which is exactly half the area of the rectangle, i.e. 16 cm².

$$\text{Area of triangle} = \tfrac{1}{2}(8 \times 4) = \tfrac{1}{2}(\text{base} \times \text{height}).$$

Example 5

Find the area of the triangle.

$$\text{Area of triangle} = \tfrac{1}{2}(3 \times 6) = \tfrac{1}{2}(18) = 9 \text{ cm}^2.$$

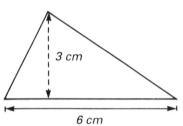

Example 6

Find the area of the triangle.

Measure the length of the base. Measure the vertical height from the base to the highest point of the triangle. Calculate

$$\text{area} = \tfrac{1}{2}(\text{base} \times \text{height}).$$

EXERCISE 13.6

Find the area of each triangle.

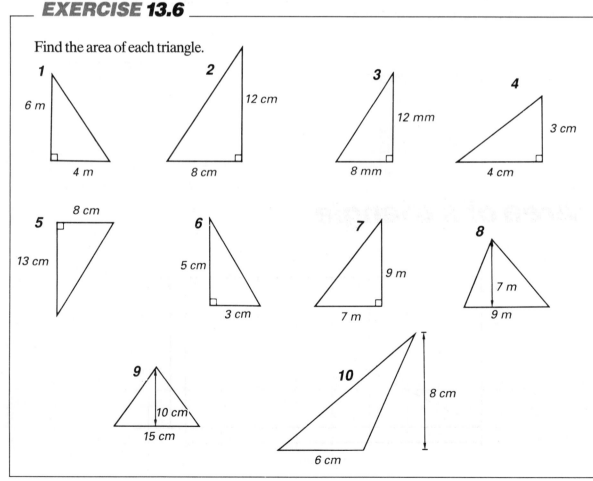

EXERCISE **13.7**

1 Find the area of a triangle whose height is 10 cm and base is 8 cm.

2 Find the area of a triangle which has a base of 9 m and a height of 16 m.

3 What is the area of a triangle which has a base of 30 cm and a height of 15 cm?

4 Find the area of a triangle which has a base of 150 mm and a height of 84 mm.

Find the area of each triangle.

5

6
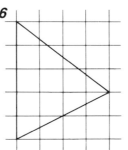

7
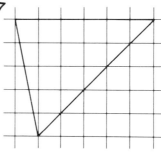

8

9
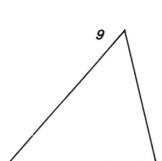

10

Investigation G

This rectangle has an area of 6 cm². How many other shapes made out of square centimetres can you find with an area of 6 cm²? Try the same investigation for 7 cm², 8 cm², etc. Could you use triangles as well?

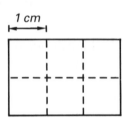

Area of compound shapes

To find the areas of more complicated shapes we usually split them up into two or more shapes whose areas we can find more easily.

Example 7

Find the total area of this shape.

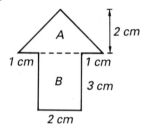

Triangle A: area = ½(base × height) = ½(4 × 2) = 4 cm²
Rectangle B: area = length × breadth = 2 × 3   = 6 cm²
                                   Total area = 10 cm²

Example 8

Find the total area, and the perimeter, of this shape.

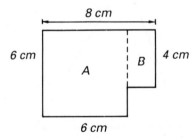

The two missing lengths as shown are 2 cm due to the measurements on the opposite sides of the shape: 8 − 6 = 2; 6 − 4 = 2.

Square A: area = 6 × 6 = 36 cm²
Rectangle B: area = 4 × 2 = 8 cm²
             Total area = 44 cm²

Perimeter: 8 + 4 + 2 + 2 + 6 + 6 = 28 cm

EXERCISE 13.8

For each shape find the total area.

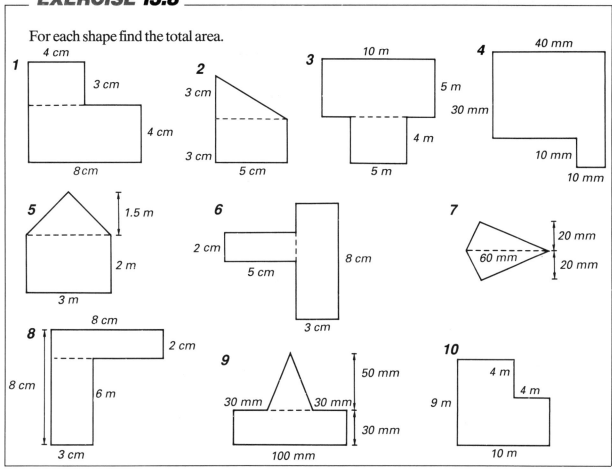

EXERCISE 13.9

For each shape find (*a*) the total area, (*b*) the perimeter.

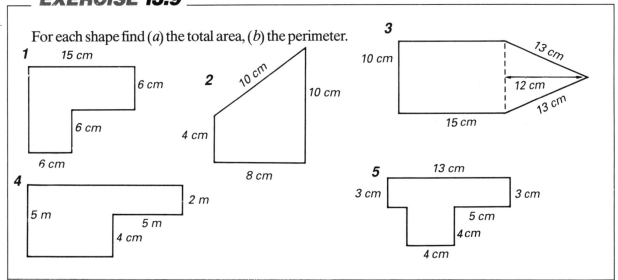

Enlargement and areas

Investigation H

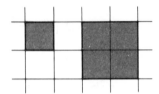

This diagram has been increased in dimensions by a scale factor 2. How many times will the small square fit into the enlarged square?

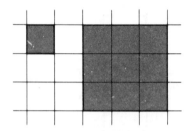

How many times will the small square fit into the enlarged square after an enlargement of scale factor 3?

Continue the experiment yourself for scale factors 4, 5 and 6, putting your results into a table:

Scale factor of enlargement	2	3	4	5	6
Small squares in enlarged square					

What do you notice about the pattern of the numbers? The number of times the small square will fit in the enlarged square is called the **area scale factor**. Can you predict the area scale factor when the scale factor for enlargement is 7 or 8?

EXERCISE 13.10

1 Repeat the investigation described above for the object shown.

Write your results into a table as before. Are the results the same as in Investigation H?

For the shapes below draw the enlargement, and write down (*a*) the scale factor, (*b*) the area scale factor of enlargement.

2
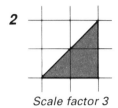
Scale factor 3

3
Scale factor 4

4
Scale factor 2

For questions 5 – 7 draw a sketch of the enlargement, writing in the new dimensions, and then write down (*a*) the scale factor, (*b*) the area scale factor of enlargement.

5
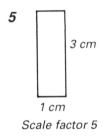
3 cm

1 cm

Scale factor 5

6
3 cm

2 cm

Scale factor 6

7
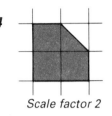
2 cm

3 cm   4 cm

1 cm

Scale factor 7

For questions 8 – 10 find (*a*) the scale factor, (*b*) the area scale factor of enlargement.

8
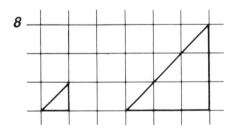

9
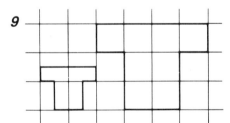

10
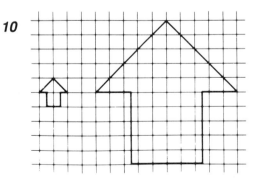

The circle

The **diameter** is measured all the way across the circle, through the centre. The **radius** is measured only half way across the circle to the centre.

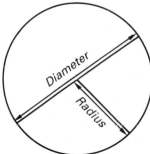

Use your ruler to measure both the radius and the diameter in this diagram.

EXERCISE 13.11

Measure (*a*) the diameter, (*b*) the radius of each circle.

1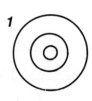

2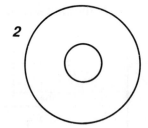

3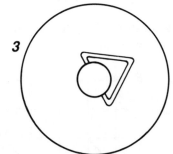

4

5

6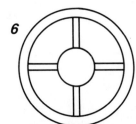

EXERCISE 13.12

Write down the length of the diameter of each circle.

 1 — 3 cm

 2 — 5 cm

 3 — 10 mm

 4 — 36 mm

 5 — 5 mm

Write down the length of the radius of each circle.

 6 — 4 cm

 7 — 16 mm

 8 — 32 mm

 9 — 7 cm

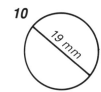 10 — 19 mm

Circumference

The **circumference** of a circle is the distance all the way around, that is, the perimeter of the circle. Try to measure the circumference of this circle. Why is it more difficult than measuring the diameter? In Book 2 you found that the circumference of a circle is slightly more than 3 times as long as the diameter. The actual figure we use is not 3, but a decimal number we know as π. Most calculators have a special button for π: if yours has, press the key, and you will see what π looks like as a number. We never have to remember the value of π since most calculators remember it for us. If you have a calculator without π, use π = 3.142. Write it down clearly in your book so you can use this value in any calculations.

We can now work out the circumference as:

circumference = π × diameter

Example 9

Find the circumference of this circle.

Measure the diameter: 2 cm
Circumference = π × diameter = 3.142 × 2 cm = 6.284 cm

Example 10

Find the circumference of this circle.

The radius is 5 cm. We need the diameter.

> Diameter $= 2 \times$ radius $= 2 \times 5 = 10$ cm.
> Circumference $= \pi \times$ diameter $= \pi \times 10 = 31.416$ cm.

Notice in the above examples the answers have been rounded after the 3rd decimal place since usually we don't need answers more accurate than this.

EXERCISE 13.13

1-6 Go back to Exercise 13.11 and find the circumference of each of the circles in questions 1–6.

7 Find the circumference of a circle of radius (*a*) 4 cm (*b*) 6 cm (*c*) 12 m (*d*) 15 mm (*e*) 30 mm (*f*) 7.5 cm (*g*) 1.5 m (*h*) 9.4 cm (*i*) 212 mm.

8 Find the circumference of a circle of diameter (*a*) 14 cm (*b*) 32 m (*c*) 6.8m (*d*) 8.4 m (*e*) 88 cm (*f*) 320 mm (*g*) 25.6 m (*h*) 54 cm (*i*) 14.8 m.

9 A bicycle wheel has a diameter of 23 inches. How far will it move the bike with each complete turn?

10 A circular lampshade of radius 15 cm is being recovered. What length of fabric is needed to go all the way round?

11 The hot-water tank in a house needs a new cover. The diameter of the cylindrical tank is 80 cm. What length of insulation is needed to cover the tank?

12 A car wheel has a diameter of 70 cm. How far will it travel on each complete turn?

13 A hose pipe reel has a radius of 30 cm. How many turns are necessary to reel in a pipe of length 1500 cm?

14 The hand of a clock is 12 cm long. How far does the tip of the hand travel in each revolution?

Cuboids

A cuboid is a 6-sided box.

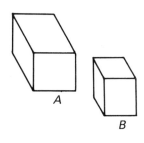

Copy the drawings into your book. It can be quite difficult to draw
3-dimensional objects on paper. It is a little easier if you have isometric
paper.

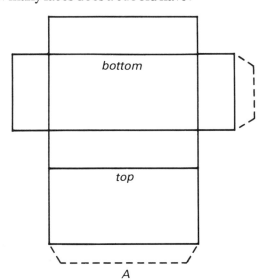

    You need some imagination to think about 3-dimensional objects.
How many faces does a cuboid have?

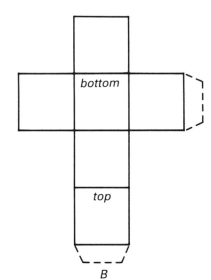

Copy the two nets on to card. Cut them out and assemble them to give
you two models of the cuboids. Why is one much wider than the other?
What effect would it have on the model if you made the net even wider?

Example 11

Find the area of each of the rectangles in the net for cuboid A.
    This is called finding the **surface area** of the cuboid.
    Find the area of each of the rectangles, add them up to find the total
area of the net. Repeat the exercise for cuboid B.

EXERCISE 13.14

For each cuboid, copy the diagram and draw its net. Use the
diagram of the cuboid to help you put the measurements on the net,
and then calculate the surface area. The net for the first question has
been drawn for you. (All measurements are in cm.)

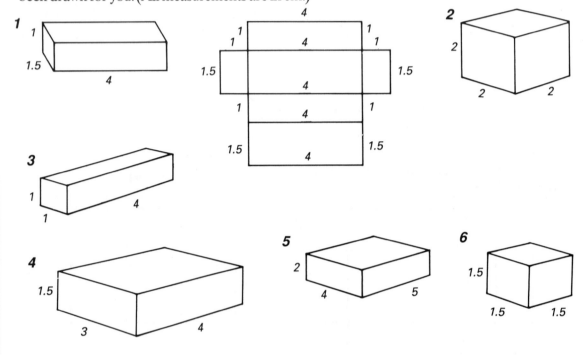

We could also find the surface area by thinking of the diagram as a solid
object.

Example 12

Find the surface area and the volume of this cuboid.

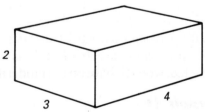

Area of top $= 3 \times 4 = 12$ cm^2. Area of bottom (same) $= 12$ cm^2.
Area of short side $= 2 \times 3 = 6$ cm^2. Other short side $= 6$ cm^2.
Area of long side $= 4 \times 2 = 8$ cm^2. Other long side $= 8$ cm^2.

Total surface area $= 8 + 8 + 6 + 6 + 12 + 12 = 52$ cm^2.

Volume $= 2 \times 3 \times 4 = 24$ cm^3.

EXERCISE 13.15

Find the surface area of each cuboid.

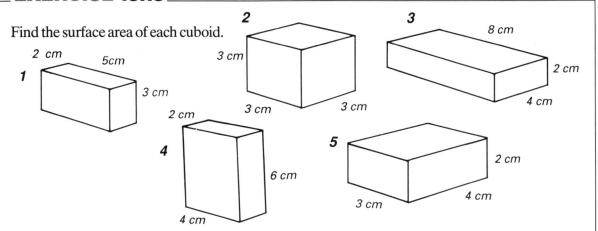

Find (*a*) the surface area, (*b*) the volume of each cuboid.

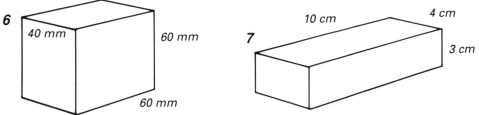

8 A box which has a base of dimensions 2 m by 3 m, and a height of 1 m.

9 A crate which measures 4 m by 3 m by 3 m.

10 A carton which has the dimensions 80 cm by 65 cm by 50 cm.

Investigation I

Cut a square of dimensions 10 cm × 10 cm out of squared paper. Cut 1 cm² off each corner, and fold the sides up to make an open box.
(*a*) What is the surface area of the box? (*b*) What is the volume of the box?

By repeating the experiment and cutting off differently sized squares, try to find the box with the maximum volume.

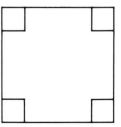

Investigation J

A small rectangular pond is to be made surrounded by a path.

Length of pond	Width of pond	Paving stones
5	4	22

Draw some differently sized ponds yourself and find the number of paving stones required for each of your drawings. Copy the table and enter your results into it. Continue this exercise until you find the pattern in the numbers to help you fill in further values in the table without having to draw diagrams.

Extension

Consider a path two or three paving stones wide around the pond, or more than one colour of paving stone which make up a pattern.

14. Probability

'What is the probability of getting a "head" when a coin is spun?' Answers are likely to vary: 'evens', 'the same as tails', 'half and half', 'one in two', 'fifty-fifty' are some replies.

All the answers are intended to mean the same thing. It is helpful, therefore, to have a clear method of representing probabilities in mathematical terms.

One of the most straightforward methods is to represent a probability by a *fraction*; the *top* figure is the number of ways a 'head' can occur (one, in this case), and the *bottom* figure is the total number of possible results that there could be (two, in this case).

$$\text{Probability of a 'head'} = \frac{1}{2} \begin{array}{l} \leftarrow \text{(total number of 'heads')} \\ \leftarrow \text{(total possible results)} \end{array}$$

Example 1

In my inside pocket I have 2 red pens and 3 blue pens. What is the probablity of picking a red pen, when I pull one pen out of my pocket?

From our statement, we need a *fraction*, with the top being the number of *red* pens and the bottom being the *total number of pens.*

$$\text{Probability of a } red \text{ pen} = \frac{\text{number of red pens}}{\text{total number of pens}} = \frac{2}{5}$$

Example 2

In a box of 10 chocolates, all looking the same, there are 3 soft centres and 7 hard centres. If you are invited to pick one, what is the probability that you pick a hard-centred chocolate?

$$\text{Probability of a hard centre} = \frac{\text{number of hard centres}}{\text{total number of chocolates}} = \frac{7}{10}$$

EXERCISE 14.1

1 In a drawer there are six handkerchiefs; five are white and one is blue. If one is selected at random, what is the probability that it is white?

2 I have a 5p piece and three 2p pieces in my pocket. If one coin is picked out, what is the probability that it is the 5p piece?

3 From a group of five girls and three boys, one person is chosen as leader. If they all have an equal chance of being chosen, what is the probability that the leader is (*a*) a girl, (*b*) a boy?

4 On my bingo card there are 15 numbers, and there are 90 numbers altogether to be called. What is the probability that the first number called is on my card? Simplify your answer. (If you aren't sure about simplifying fractions, look back to Chapter 10.)

	23		42	51	61		82
2		39		55	63	79	
	16	25		43	59		88

5 After Saturday's cricket match, in which five of our team were caught out, one of the eleven players was chosen, at random, to collect the match fees. What is the probability that this player was one of those who was caught out?

6 Manju has a box of twenty cassette tapes, seven of which are by the Beatles (the tapes were her grandmother's!). If she takes a tape out of the box, what is the probability that it is *not* a Beatles' tape?

7 Seven letters arrive through the post, addressed to 'M. Henderson'. If three are for Mike and four are for Michelle, what is the probability that the first one opened is for Mike?

8 One pupil is selected at random from a class of 16 boys and 12 girls. What is the probability that the selected pupil is a girl?

9 On a stall at a School Fayre, a tin contains 100 straws with the hidden ends painted in different colours. If you know that there are 15 green straws, what is the probability that a straw picked out is green?

10 Of the nine pairs of tights in her cupboard, Mandy knows that five pairs have ladders in them. When she picks a pair from her cupboard, what is the probability that the pair she picks has no ladder?

Dice and cards

Because most people are familiar with dice and playing cards, and can actually handle them, there are many examples in probability that use these. In fact, the whole theory of probability was derived from studying the predicted results when dice were thrown. Experienced card players can improve their game if they study the probabilities of certain combinations of cards.

Example 3

When a die is thrown, what is the probability of getting (*a*) a six, (*b*) an even number, (*c*) a number other than three, (*d*) either a three or a four, (*e*) a seven?

There are six possibilities (1,2,3,4,5 or 6), each one equally likely.

(*a*) Probability of a six $= \frac{1}{6}$

(*b*) Probability of an even number $= \dfrac{\text{number of even numbers}}{6}$

$$= \frac{3}{6}$$

(*c*) Probability of number other than three $= \dfrac{\text{number of 'not threes'}}{6}$

$$= \frac{5}{6}$$

(*d*) Probability of either a three or a four $= \frac{2}{6}$

(*e*) Probability of a seven $= \dfrac{\text{number of sevens}}{6}$

$$= \frac{0}{6}, \quad \text{i.e. } 0$$

EXERCISE 14.2

1 When a die is thrown, what is the probability of getting (*a*) a four, (*b*) a number greater than four?

2 In a game of Snakes and Ladders, I need either a 1 or a 2, or else I can't move. What is the probability that I can move when I throw?

3 In another game, an odd number will enable me to move on my next turn. What is the probability that I can move on my next turn?

4 As long as I don't get a 4 or a 5 I am 'safe'. What is the probability of remaining 'safe' after my next throw?

5 I need to throw a three to win a game of 'Beetle'. What is the probability that I win with this throw?

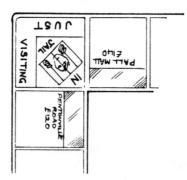

6 If I throw a 2, I have to 'go to jail'. What is the probability that I do not go to jail?

7 I win if my total is either 10 or 11, when I throw a die twice. What is the probability that I win, if my first throw is (*a*) a 6, (*b*) a 4 ?

8 I can move if my score is 3 or more. What is the probability that on my next throw I can move?

With a pack of cards, there are many possible arrangements. First, let us make sure that we know the composition of a pack of cards.

There are four **suits**: clubs, diamonds, hearts and spades. Clubs and spades are black, diamonds and hearts are red.

In each suit there are 13 cards: ace, 2, 3, 4, 5, 6, 7, 8, 9, 10, jack, queen and king.

So altogether there are 4 × 13 = 52 cards in a pack. (Some packs have extra cards called jokers. We are ignoring these.)

Example 4

I pick a card from a pack. What is the probability that it is (*a*) a red card, (*b*) a five, (*c*) the ace of spades, (*d*) not a diamond, (*e*) a black jack?

(*a*) Probability of a red card $= \frac{26}{52}$ ($= \frac{1}{2}$), as there are 26 red cards (all the diamonds and hearts).

(*b*) Probability of a five $= \frac{4}{52}$ ($= \frac{1}{13}$), as there are 4 fives, one in each suit.

(*c*) Probability of the ace of spades $= \frac{1}{52}$, as there is only one ace of spades.

(*d*) Probability of 'not a diamond' $= \frac{39}{52}$ ($= \frac{3}{4}$), as there are 39 cards which are not diamonds.

(*e*) Probability of a black jack $= \frac{2}{52}$ ($= \frac{1}{26}$), as there are two black jacks, the jack of clubs and the jack of spades.

Investigation A

(*a*) Shuffle a pack of cards, and keep turning over the cards one by one until you turn up the first heart (or club, or spade or diamond – choose whichever suit you please). Note down the number of cards that you needed to turn over.

(*b*) Shuffle, and repeat, noting again the number of cards turned up before a heart appears.

(*c*) Repeat the process 50 times.

(*d*) Estimate the probability of turning up the first heart (i) with the first card, (ii) with the second card, (iii) with the third card, etc.

EXERCISE 14.3

1 I am dealt a card from a pack. What is the probability that it is (*a*) a king, (*b*) a spade, (*c*) a two, three or four?

2 From a pack, I pick a card. What is the probability that it is (*a*) a picture card (i.e. a jack, queen or king), (*b*) a club?

3 In dealing cards from a pack, I accidentally drop a card on the floor. What is the probability that it is (*a*) the six of hearts, (*b*) any heart, (*c*) any ace?

4 From a pack I select the thirteen hearts, shuffle them and turn the top card over. What is the probability that it is (*a*) the four, (*b*) the nine, (*c*) an even number?

5 I have an old pack of cards from which the five and six of diamonds have been lost, leaving 50 cards. What is the probability that, if I pick one card, it will be (*a*) a black card, (*b*) a seven, (*c*) a diamond, (*d*) a six?

6 The seven cards I am dealt contain four clubs, two diamonds and a spade. My partner picks one of these cards, without looking. What is the probability that it is (*a*) a club, (*b*) a diamond, (*c*) a heart?

7 What is the probability that the first card I am dealt is (*a*) a six, (*b*) a diamond, (*c*) the six of diamonds, (*d*) either a nine or a ten?

8 I have been dealt four cards, and all are hearts. What is the probability that the next card I am dealt is also a heart?

9 All the twelve picture cards (i.e. jacks, queens and kings) are shuffled, and one is turned up. What is the probability that it is (*a*) a spade, (*b*) a queen, (*c*) either a king or a club?

10 I select three cards from a pack. When I turn over the first two, they are both kings. What is the probability that the third card is also a king?

Investigation B

If I have two cards (say an ace and a two), I can play them in two ways: either the ace first followed by the two, or the two first followed by the ace.

Investigate the number of different ways of playing three cards.

Extend to four, five and six cards.

See if you can find a pattern, or rule, which would enable you to calculate the number of different ways of playing ten cards, or thirteen, or any number.

Investigation C

In the game of Poker, a 'run' is a sequence of five consecutive cards. For example, '5, 6, 7, 8, 9', '8, 9, 10, J, Q' and 'A, K, Q, J, 10' are all 'runs'.

Assume that you have four cards. Investigate the probability that your fifth card completes the run, if your four cards are (*a*) 3, 4, 5 and 6, (*b*) 8, 9, 10 and J, (*c*) 4, 6, 7 and 8, (*d*) J, Q, K and A, (*e*) 6, 7, 8 and 10.

Extension

Choose all sets of four cards which give the possibility of completing a run, and investigate the probability of making it in each case.

Impossibility and certainty

Example 5

In a box there are 3 red, 4 green and 5 white buttons. One is selected at random. Find the probability that it is (*a*) green, (*b*) white, (*c*) not green, (*d*) either red or green, (*e*) yellow, (*f*) either red, green or white.

(*a*) Probability of a green button $= \frac{4}{12}$, as there are 4 greens in a total of 12.
(*b*) Probability of a white button $= \frac{5}{12}$ (there are 5 white buttons).
(*c*) Probability of a button not being green $= \frac{8}{12}$, because there are 8 buttons not green (the 3 reds and the 5 whites).
(*d*) Probability of either a red or green button $= \frac{7}{12}$ (there are 3 red and 4 green buttons).
(*e*) Probability of a yellow button $= \frac{0}{12}$, or 0, as there are no yellow buttons. It is thus *impossible* to pick out a yellow button.

A probability of 0 means that an event can't happen – you can't pick a yellow button if there aren't any!

A probability of 0 means 'impossibility'

(*f*) Probability of a red, green or white button $= \frac{12}{12}$, or 1, as the button *has* to be red, green or white. These are the only colours possible.

A probability of 1 means that an event *must* happen.

A probability of 1 means 'certainty'

It follows that a *small* probability (unlikely to happen) is a *small* fraction, e.g. $\frac{1}{10}$, or $\frac{3}{50}$.

Similarly, a *high* probability (likely to happen) is a *large* fraction, e.g. $\frac{9}{10}$, or $\frac{47}{50}$.

EXERCISE 14.4

1 I am dealt a hand of five cards; the seven of clubs, the ten of clubs, the queen of diamonds, the queen of spades and the six of spades. My partner picks one of these cards at random. What is the probability that it is (*a*) the six of spades, (*b*) a queen, (*c*) not a diamond, (*d*) a king?

2 In a game, I have to close my eyes and pick a straw from a bunch of twenty. Eight of them are white, three are black and the remaining nine are red. What is the probability that I pick a straw which is (*a*) black, (*b*) not red, (*c*) either red or white, (*d*) green, (*e*) either black, red or white?

3 There are twelve cans of fizzy drink in a cardboard box; five are of lemonade, three are orangeade and four are cherryade. One is picked from the box at random. What is the probability that it is (*a*) cherryade, (*b*) not lemonade, (*c*) coke?

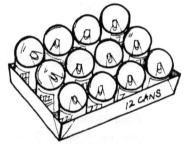

4 Two boys, Steven and Stan, and three girls, Shajada, Shirley and Sue, are playing a game. What is the probability that the winner is (*a*) a boy, (*b*) Shirley, (*c*) Stella, (*d*) a person whose name begins with an 'S'?

5 A letter is selected at random from the word **abracadabra**. What is the probability that it is (*a*) an 'a', (*b*) a 'b', (*c*) an 'e'?

6 In each packet of breakfast cereal there is one model animal. There are six animals in the set; a lion, an elephant, a rhinoceros, a tiger, a giraffe and a monkey. What is the probability that, in your next packet, the animal is (*a*) a tiger, (*b*) either a giraffe or a monkey, (*c*) not a lion, (*d*) a hippopotamus?

7 In my money box I have three 50p coins, six 10p coins, four 5p coins and three 2p coins. How much is this altogether?
  I shake the money box and one coin drops out. What is the probability that it is (*a*) a 5p coin, (*b*) a 50p coin, (*c*) a 20p coin, (*d*) less than £1?

8 There are 14 girls and 11 boys in class 3BH. Joanne, one of the girls in 3BH, is carrying their 25 exercise books when one of them falls off. What is the probability that it belongs to (*a*) a boy, (*b*) Joanne, (*c*) someone in 3BK?

9 Mrs Wilson brings in a tray with twelve cups of tea for the ladies at the WI Committee meeting. She has forgotten, however, into which of the five cups she has put sugar. What is the probability that the first cup she takes off the tray is (*a*) sugared, (*b*) coffee, (*c*) not sugared, (*d*) tea?

10 During a game of Scrabble I have these eight letters on my rack: A, E, T, E, E, Q, A, F. As I should have only seven letters, my opponent removes one letter without looking at it. What is the probability that it is (*a*) an E, (*b*) the Q, (*c*) not an A, (*d*) not a T, E or A, (*e*) a V?

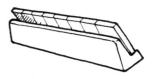

Comparing probabilities

Sometimes it is difficult to tell which probability is greater when there are two probabilities nearly equal to each other. We can compare probabilities more easily if we change each fraction into a decimal. (Refer back to Chapter 5 if you need to remind yourself how to do this.)

Example 6

Diana has twelve cards of which five are black. Peter has thirty cards of which thirteen are black. If you wanted to pick a black card, who would you choose to pick from, Diana or Peter?

The probability of picking a black card from Diana is $\frac{5}{12}$.
The probability of picking a black card from Peter is $\frac{13}{30}$.
To find out which is greater, change each fraction into a decimal

$$\frac{5}{12} = 0.416\,66\ldots \quad \text{and} \quad \frac{13}{30} = 0.433\,33\ldots$$

So the probability of black from Diana is 0.416 66 ... and the probability of black from Peter is 0.433 33 ....

It is better, therefore, if you want a black card, to pick a card from Peter, as the probability is higher (just) than from Diana.

EXERCISE 14.5

1 Which is the greater probability, $\frac{1}{3}$ or $\frac{7}{10}$?

2 Which is the greater probability, $\frac{8}{23}$ or $\frac{32}{100}$?

3 Jim says that a probability of $\frac{6}{10}$ is better than $\frac{13}{25}$. Is he correct?

4 Anne says that a probability of $\frac{6}{10}$ is better than $\frac{5}{9}$. Is she correct?

5 You are given two options: either throw a six with one throw of a die, or pick either a three or a four from a pack of cards. Which option would you go for? Show why you chose your answer.

6 Last year $\frac{4}{5}$ of the year group went on the school trip. This year there were 143 on the trip, out of a total of 187. Was it more likely that someone went on the school trip last year or this year?

7 'Your chance of winning the raffle is less than $\frac{1}{50}$', I am told. If I have bought 12 raffle tickets, and there are 500 tickets sold, what is the probability that I will win? Have I been told the truth?

8 Eight of the twenty-six pupils in class 3BJ wear glasses, while in class 3BR there are six out of nineteen who wear glasses. Which class is the more likely to have someone who wears glasses?

9 There are 7 red counters and 13 blue counters in a white bag. In a black bag there are 11 red counters and 25 blue counters. If you are allowed only one bag from which you hope to pick a red counter, would you choose the white bag or the black one?

10 In the fourth year there are 178 pupils, and 33 of them do Media Studies. In the fifth year there are 207 pupils, of whom 38 do Media Studies. One pupil from each year group is selected at random. Work out whether or not the fifth-year pupil is more likely to be doing Media Studies than the fourth-year pupil.

Revision exercises: Chapters 10-14

1 47 mm + 39 mm

2 94 mm + 26 mm

3 17.4 cm + 30.6 cm

4 243 mm + 628 mm

5 6 cm + 15.4 cm

6 67.8 m + 42.8 m

7 890 mm + 980 mm

8 27 mm + 27 mm + 64 mm

9 3.73 m + 5.25 m + 3 m

10 (5 × 210 mm) + 400 mm

11 95 cm − 66 cm

12 2.47 km − 1.81 km

13 52.2 m − 37.6 m

14 12.6 m − 7.35 m

15 206 mm − 54 mm

16 6.83 km − 5.92 km

17 352 m − 177 m

18 17.4 cm − 8.03 cm

19 300 km − 74.6 km

20 (3 × 607 m) − 1264 m

21 Along the A40, it is 153 miles from Fishguard to Gloucester, and a further 109 miles from Gloucester to London. How far is it from Fishguard to London, along the A40?

22 Three ropes are 17.4 m, 13.7 m and 25.8 m long. What is the total length of all three ropes?

23 To stop people walking across the wicket, a cricket groundsman erects a rectangle of ropes around it. The rectangle measures 31 m by 46 m. Would a rope of length 140 m be long enough to make the rectangle to go round the wicket?

24 I need 270 mm of string to make a Christmas card. What length of string will I need for 8 cards? Give your answer in millimetres, then in centimetres, then in metres.

25 From a plank of wood 2 m long I cut off 87 cm. What length is left?

26 If I cut three pieces, each 37 cm, from a 2 metre plank, how much of the plank is left?

27 How many pieces, each 37 cm long, can I cut from a plank of wood 2 m long? How long is the piece left over?

28 I need to drive from Bristol to Leeds, a distance of 312 km. After two hours I have travelled 158 km. How much further have I to go?

29 The height of my bedroom is 2.92 m from floor to ceiling. What length of wallpaper will I need between the ceiling and the top of the door frame, if the door frame is 1.97 m from the floor?

30 For a bookcase, I need two pieces of wood 975 mm long, and three pieces of wood 855 mm long. What is the total length of wood that I require to make the bookcase?

31 $47\,g + 32\,g$

32 $632\,g + 151\,g$

33 $17.9\,g + 28.4\,g + 22.4\,g$

34 $2.43\,kg + 3.51\,kg$

35 $4.67\,kg + 1.52\,kg - 3.54\,kg$

36 $274\,g - 134\,g$

37 $34.7\,g - 7.89\,g$

38 $345\,g - 67.8\,g$

39 $254.2\,kg - 79.42\,kg$

40 $1.64\,kg - 860\,g$

41 I mix 250 g of flour with 124 g of margarine. What weight of mixture do I have now?

42 Five boxes each weigh 13.7 kg. What is the total weight of all five boxes?

43 A lift will take a maximum of 600 kg, otherwise it will not operate. Five people enter the lift. Their weights are 124 kg, 103 kg, 163 kg, 142 kg and 137 kg. Will the lift be able to operate?

44 In a weightlifting competition, a Russian lifted 236.5 kg, 180 kg and 237.5 kg. What was the total weight lifted?

45 I buy two jars of jam, each weighing 340 g, and three tins of meat, each weighing 397 g. What is the total weight of these five items, in kilograms?

46 What is the total weight of a team of four wrestlers weighing 83.5 kg, 94.5 kg, 79.3 kg and 72.2 kg?

47 My wheelbarrow can carry about 90 kg of sand. How many trips will I need to make if I have to move a heap of sand weighing 650 kg?

48 I melt a 500 g bar of cooking chocolate, and make six moulds, each weighing 42 g. How many more moulds, each weighing 76 g, can I make from the remaining chocolate?

49 Can I make twelve cakes, each weighing 275 g, from a bowl containing 3.15 kg of cake mixture?

50 Eight cartons, each containing 263 g of cream, are put on the scales. Their total weight is 2.24 kg. How heavy is one empty carton?

51 75 ml + 235 ml

52 638 ml + 492 ml

53 6.75 litres + 7.03 litres

54 12.6 litres + 34.5 litres

55 60 ml + 83 ml + 95 ml

56 300 litres − 164.5 litres

57 3.63 ml − 1.19 ml

58 69.5 litres − 30.8 litres

59 342 ml + 427 ml − 503 ml

60 2.3 litres − 478 ml

61 A garage has a full 35 litre drum of antifreeze, plus another 35 litre drum from which 23 litres have been poured out. How many litres of antifreeze does the garage still have?

62 I have to mix 56.2 ml of one chemical with 85.6 ml of another chemical. How many millilitres have I altogether?

63 I have a set of wine glasses. Each one can hold 14 cl. How many glasses can I pour from a 75 cl bottle of wine?

64 The petrol tank on my car will hold 31.5 litres when full. Before starting out on a journey I call at a petrol station and fill the tank, by adding 23.7 litres of petrol. How much petrol must there have been in the tank already?

65 My favourite washing-up liquid comes in three sizes: 550 ml, 870 ml, and 1.24 litres. If I buy one of each, how much washing-up liquid will I have? Give your answer in millilitres, then in litres.

66 In a science experiment I run off 53.5 ml of acid into a flask which can hold 250 ml. How many times can I do the experiment without having to empty the flask?

67 The urn used for making tea for a cricket match holds 25 litres. The cups can each hold 280 ml. Will there be enough tea for both teams, plus umpires (i.e. 24 altogether) to have two cups of tea each?

68 If I need to fill a 1.5 litre bottle, using a jug which holds 230 ml. How many full jugs will I need?

69 During a week my milkman delivers sixteen pints of milk. How many litres is this, if 1 litre = 1.76 pints?

70 A plastic container is advertised as being able to hold 5 gallons of beer or 25 litres of wine. Which is the greater amount? You will need this information: 1 gallon = 8 pints; 1 litre = 1.76 pints.

71 $\frac{2}{5}+\frac{1}{5}$ **72** $\frac{3}{8}+\frac{2}{8}$

73 $\frac{3}{10}+\frac{4}{10}$ **74** $\frac{1}{6}+\frac{4}{6}$

75 $\frac{5}{16}+\frac{5}{16}+\frac{1}{16}$ **76** $\frac{4}{5}+\frac{3}{5}$

77 $\frac{3}{4}+\frac{3}{4}+\frac{3}{4}$ **78** $\frac{6}{7}+\frac{4}{7}+\frac{4}{7}$

79 $\frac{7}{12}+\frac{11}{12}$ **80** $\frac{5}{8}+\frac{6}{8}+\frac{7}{8}$

81 $\frac{5}{6}-\frac{2}{6}$ **82** $\frac{7}{10}-\frac{1}{10}$

83 $\frac{13}{20}-\frac{4}{20}$ **84** $\frac{7}{9}-\frac{2}{9}$

85 $\frac{3}{8}+\frac{7}{8}-\frac{5}{8}$ **86** $\frac{4}{5}+\frac{2}{5}-\frac{3}{5}$

87 $\frac{79}{100}-\frac{24}{100}$ **88** $\frac{7}{12}-\frac{9}{12}+\frac{4}{12}$

89 $\frac{15}{16}-\frac{3}{16}$ **90** $\frac{5}{9}+\frac{3}{9}-\frac{8}{9}$

91 $\frac{1}{6}+\frac{1}{3}$ **92** $\frac{3}{10}+\frac{1}{5}$

93 $\frac{7}{8}-\frac{1}{4}$ **94** $\frac{1}{4}-\frac{1}{6}$

95 $\frac{5}{12}+\frac{5}{6}$ **96** $\frac{1}{2}+\frac{1}{3}+\frac{1}{4}$

97 $\frac{17}{20}-\frac{2}{5}$ **98** $\frac{1}{5}-\frac{1}{6}$

99 $\frac{9}{10}+\frac{2}{3}$ **100** $\frac{3}{50}+\frac{7}{30}$

101 If my car can average about 47 miles on a gallon of petrol on a long run, about how many gallons will I need to drive the 235 miles from Edinburgh to Sheffield?

102 If a litre of paint covers about 5 square metres, how many 5-litre tins will I need to paint a wall which has an area of 120 square metres?

103 How much, to the nearest £, will it cost for eight pairs of sheets at £13.95 a pair?

104 About how many litres of petrol, at 37.6p per litre, can I buy for £10.00?

105 How much money, to the nearest £100, will have been taken at a turnstile through which 403 people passed, each paying £6.50?

106 It costs £14.89 for me to stay at a hotel for one night. How much will it cost three of us to stay for four nights at the hotel? Give your answer correct to the nearest £10.

107 If my rate of pay is £4.49 an hour, what are my earnings for a week of 39 hours? Give your answer correct to the nearest £20.

108 The fee to join our local golf club is £130 a year. If I expect to play 27 rounds of golf in a year, about how much does this cost me for one round, to the nearest £?

109 My daily bus fare to school is 16p each way. About how much bus fare do I pay in a year of 39 five-day school weeks, correct to the nearest £?

110 A bag of coal costs £5.32. How much will I pay for coal in a year in which I buy 34 bags? Give your answer to the nearest £10.

In questions 111–115 write down the reading on the dials.

116 Write down these times as they would appear on a digital watch.

(a)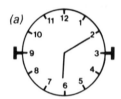

(b)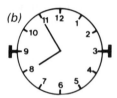

117 Draw two watch faces, and indicate these times.

(a)

(b)

118 How many times will the last digit of this counter change before the first digit 8 becomes a 9?

119 Write down the reading on this scale.

120 Estimate these readings.

(a)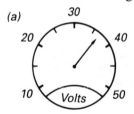

(b)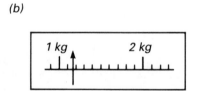

121 Using the diagram of a thermometer, what is (a) 55°F in °C, (b) 8°C in °F?

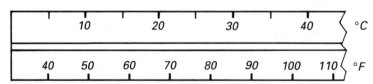

122 From the diagram, write (a) 2.5 cm in inches, (b) $1\frac{1}{4}$ inches in centimetres.

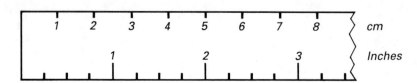

123 Use the graph to change (*a*) 2 stones into kg, (*b*) 10 kg into stones.

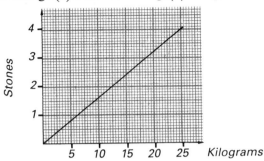

124 Use the graph to change (*a*) 3 litres into pints, (*b*) 2.5 pints into litres.

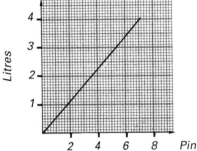

125 How many degrees are there between W and NE?

126 How many degrees are there between SW and SE?

127 How many degrees are there between NW and E?

128 If you are facing NE and turn 135° clockwise, in what direction would you be facing?

129 If you are facing SE and turn 270° anticlockwise, in what direction would you be facing?

130 Use a protractor to draw an angle of (*a*) 55° (*b*) 45° (*c*) 95° (*d*) 230° (*e*) 310° (*f*) 285°.

131 Measure the angles in each of these triangles.

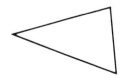 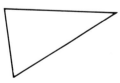

132 Draw a North line, and then use a protractor to draw a bearing of (*a*) 080° (*b*) 250°. (*c*) 125° (*d*) 320° (*e*) 195° (*f*) 305°.

133 What is the bearing of P from O in each of these diagrams?

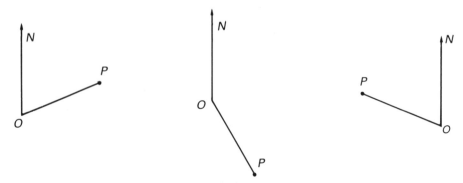

134 A pilot flies from the airport A on a bearing of 055° for a distance of 450 km over Barton Hill. He then changes course, to a bearing of 130°, for 340 km. Draw a scale diagram of the journey, and find out the distance and bearing of the aeroplane from the airport after the second leg.

135 Draw accurate scale diagrams of these sketches of journeys.

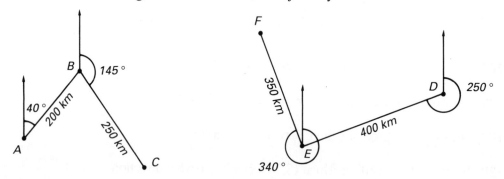

136 A plane starts on a bearing of 250° from Ripley. It flies for 430 km, then takes a new bearing of 90° and flies for a further 330 km to Whitefield. Find the distance and bearing of Whitefield from Ripley.

137 I walk 15 km in 3 hours. What is my average walking speed?

138 I cycle the 5 km to school in 15 minutes. What is my average cycling speed, in km/h?

139 If I drive the 120 miles to my aunt's house in $2\frac{1}{2}$ hours, what is my average speed?

140 An aeroplane takes off from Heathrow airport at 10.00 a.m. and lands at Rome four hours later, having flown 848 miles. What is the plane's average speed, in m.p.h.?

141 Six friends have a meal at the County Hotel. The bill comes to £44.40. What is the average price of a meal?

142 I count the number of potatoes in nine bags, with the following results: 13, 15, 13, 18, 16, 17, 15, 10 and 18. What is the average number of potatoes per bag?

143 My telephone bills for the last four quarters have been £68.35, £73.61, £50.46 and £94.06. What was my average quarterly bill?

144 A dozen (12) eggs were weighed individually. Their weights were 58 g, 64 g, 57 g, 52 g, 60 g, 48 g, 54 g, 53 g, 59 g, 51 g, 55 g and 61 g. What is the average weight of one of these eggs?

145 What is the average weight of a teabag in a box of 160 teabags which weighs 500 g? Give your answer to the nearest 0.1 g.

146 How far will I run in two hours if my average running speed is 9 km/h?

147 If I drive at an average 52 miles an hour, how far will I travel in $2\frac{1}{2}$ hours?

148 A factory produces 230 components an hour. How many components will it produce in a working day of 7.5 hours?

149 If I jog at an average speed of 12 km/h, how far will I go in 20 minutes?

150 'At an average speed of 85 km/h, a car will travel over 200 km in $2\frac{1}{2}$ hours.' Check to see whether or not this statement is correct.

151 How long will it take to complete a motor cycle race over a distance of 320 km at an average speed of 120 km/h? Give your answer in hours and minutes.

152 A runner averages 18 km per hour in a race of 1500 metres. How long will she take to complete the race?

153 If a long-playing record makes 33 revolutions per minute, how many revolutions will it make in playing one side of 24 minutes?

154 A machine manufactures 5700 components in a working week of 38 hours. How many components per hour is this on average?

155 One lap of a go-kart track is 200 metres. A race of 10 laps is won in a time of 4 minutes. Work out the average speed of the winner in (*a*) metres per second, (*b*) kilometres per hour.

156 A kettle takes 90 seconds to boil from a temperature of 10°C to 100°C. Work out the average rate of increase of temperature, in °C per second.

157 What distance will a train travel in 3 minutes at a speed of 140 km/h?

158 How long will it take the same train as in question 157 to cover a distance of 8.75 km?

159 Find (*a*) the perimeter, (*b*) the area, of the top of the table.

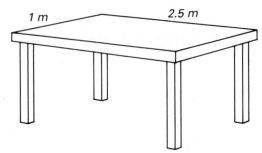

1 m   2.5 m

160 Find the perimeter and area of the square with side 3.5 cm.

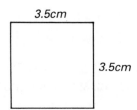

3.5cm

3.5cm

161 The perimeter of the rectangle is 28 cm. Find its area.

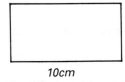

10cm

162 Find the area of a square of side (*a*) 4.5 m, (*b*) 6.2 m.

163 Work out the dimensions of another rectangle which has the same area as this one, measuring 6 cm by 4 cm.

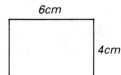

6cm

4cm

164 Find the area of this carpet for a dining room.

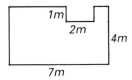

165 The carpet has to leave a border 1 m wide.

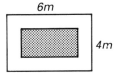

(*a*) Find the area of the carpet.

(*b*) What is the area of the border?

166 Calculate the area of the carpet required for this bedroom.

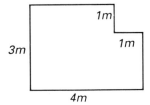

167 Find (*a*) the perimeter, (*b*) the area, of this shape.

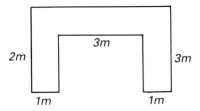

In questions 168 – 171, find the areas of the triangles.

168

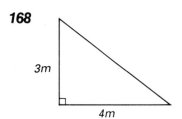

169

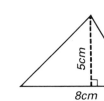

170

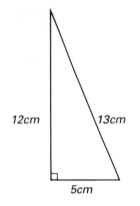

12cm   13cm

5cm

171

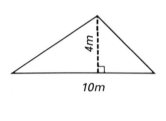

4m

10m

In questions 172–176, find the total area of each shape.

172

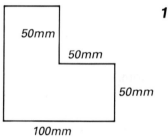

50mm

50mm

50mm

100mm

173

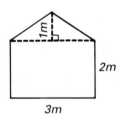

1m

2m

3m

174

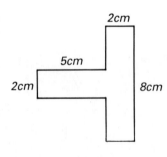

2cm

5cm

2cm   8cm

175

80mm

100mm

80mm

176

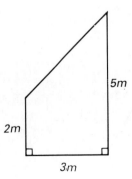

5m

2m

3m

177 Measure the diameter of this circle.

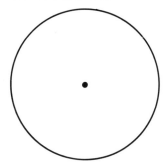

178 Measure the radius of this circle.

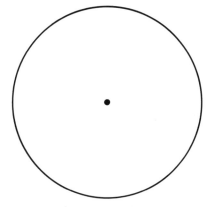

179 What are the radii of these circles?

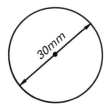
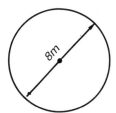

180 What are the diameters of these circles?

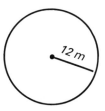

181 A circle has a radius of 5 cm. Find its circumference.

182 Calculate the circumference of a circle with a diameter of 8 cm.

183 A circular clock face has a radius of 10 cm. Calculate the circumference of the clock face.

184 A hoop has a diameter of 25 cm. What is its circumference?

185 A hot-water cylinder has a radius of 450 mm. What is the circumference of the water cylinder?

186 Find the surface area of the six faces of the box made from these shapes (ignore the flaps).

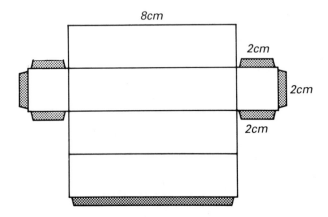

187 Find the surface area (all six faces) of each of these boxes.

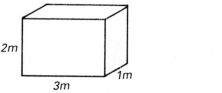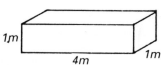

188 From a group of five girls and three boys, one is picked at random. What is the probability that it is (*a*) a girl, (*b*) a boy?

189 Seven men and five women play in a domino competition. What is the probability that the winner is (*a*) a man, (*b*) a woman?

190 One pupil in a class scored 93% in a recent maths test. If there are 12 boys and 15 girls in the class, what is the probability that the 93% mark was scored by (*a*) a girl, (*b*) a boy?

191 When a roulette wheel is spun, the ball has an equal chance of landing in any of the 37 numbers from 0 to 36. What is the probability that the ball lands on a number which is (*a*) odd, (*b*) the '0', (*c*) between 1 and 12 inclusive, (*d*) between 19 and 36 inclusive, (*e*) a multiple of 3, (*f*) 22, 23 or 24?

192 My box has 22 pens in it; 12 blue, 7 black and 3 red. If I take one out without looking at it, what is the probability that it is (*a*) black, (*b*) red, (*c*) not blue, (*d*) green?

193 When I throw a die, find the probability that I throw (*a*) a four, (*b*) an odd number, (*c*) a one or a two, (*d*) a nine.

194 I pick a card from a full pack of 52 cards. What is the probability that the card I pick is (*a*) black, (*b*) a diamond, (*c*) a 'five' (*d*) the two of hearts, (*e*) either an eight or a nine, (*f*) not a spade, (*g*) neither a king nor a queen, (*h*) either red or black?

195 Which probability is lower: $\frac{5}{7}$ or $\frac{7}{10}$?

196 Which is the higher probability: $\frac{3}{8}$ or $\frac{19}{50}$?

197 Arrange these probabilities in order, beginning with the lowest: $\frac{4}{11}$, $\frac{19}{40}$, 0.47, $\frac{5}{14}$, 45%.

198 In a jar there are 5 green and 3 red beads. These 8 beads were taken from a box which now has 27 green and 16 red beads in it. If I can pick only one bead, and I want a green one, should I pick from the jar or the box? (Show your working.)

First published 1990

Published by
MACMILLAN EDUCATION LTD
Houndmills, Basingstoke, Hampshire RG21 2XS
and London
Companies and representatives
throughout the world

Cover design by Plum Books, Southampton

Printed in Hong Kong

British Library Cataloguing in Publication Data
Miller, Jim
Macmillan secondary mathematics. Bk.3X
1. Mathematics. Questions and answers. For schools
I. Title   Newman, Graham
510'. 76
ISBN 0 333 34608 4

The authors and publishers wish to thank the following for permission
to use copyright material: Cambridge University Press for the 'Pascal'
Triangle in *Science and Civilisation in China* by Joseph Needham,
111, 135, 1959; British Railways Board for a timetable, May 1987,
and Intercity logo; Greater Manchester Passenger Transport Executive
for a timetable, May 1987; Wickes Building Supplies Ltd for material
from a catalogue, Dec. 1987.

Cover picture copyright The Image Bank

Every effort has been made to trace all the copyright holders but if
any have been inadvertently overlooked the publishers will be pleased
to make the necessary arrangement at the first opportunity.